Groundwater and Subsurface Environments

Makoto Taniguchi

Editor

Groundwater and Subsurface Environments

Human Impacts in Asian Coastal Cities

 Springer

Editor
Makoto Taniguchi, D.Sc
Professor
Research Institute for
    Humanity and Nature
457-4 Motoyama, Kamigamo
Kita-ku, Kyoto 603-8047
Japan
makoto@chikyu.ac.jp

ISBN 978-4-431-53903-2   e-ISBN 978-4-431-53904-9
DOI 10.1007/978-4-431-53904-9
Springer Tokyo Dordrecht Heidelberg London New York

Library of Congress Control Number: 2011920371

Springer is part of Springer Science+Business Media (www.springer.com)

Preface

Land subsidence due to excessive groundwater pumping, groundwater contamination, and subsurface thermal anomaly have occurred frequently in Asian coastal cities. In this volume, the relationship between the stage of a city's development and subsurface environment issues have been explored and go beyond the boundaries between surface/subsurface and land/ocean in Asian coastal cities. The results presented are the outcome of a study undertaken by the Research Institute for Humanity and Nature (RIHN) titled "Human impacts on urban subsurface environment" (project leader: Makoto Taniguchi).

Included in this work are several advanced methods that were developed by RIHN and used to evaluate subsurface environmental problems. Numerical modeling of the subsurface environment was established for Tokyo, Osaka, Bangkok, and Jakarta to evaluate the groundwater recharge rate/area, residence time, and exchange of fresh/salt water. Using updated satellite GRACE data, we have succeeded in revealing not only seasonal variations but also a secular trend in the water mass variations in the Chao Phraya River Basin. We have also developed groundwater aging methods using CFCs and ^{85}Kr. The groundwater flow system in the urban aquifer has been highly disturbed by human pumping. A dominant vertical downward flux was revealed in the urban area using CFCs and ^{14}C. The 3-D groundwater simulation (MODFLOW) showed a spatial change in the groundwater recharge area, the major recharge area of the pumped aquifer. This spatial change in the groundwater potential was strongly affected by regional groundwater pumping regulations, and the success or failure of such regulations depended mostly on the availability of alternative water resources for the city area and the legal aspects of groundwater resources.

Accumulations of trace metals in the sediment and dissolved nitrogen in groundwater in Asian cities are also discussed in this book. Various nitrogen sources and areas of denitrification were found by measuring nitrogen isotope distributions in groundwater. Groundwater salinization was found in Osaka, Bangkok, and Jakarta. The differences in marine alluvium volume (same as topographic gradient), natural recharge, and intensity of pumping period resulted in different degrees of salinization. Organic and metal pollution histories were reconstructed in Asian cities using marine sediment cores. Minor amounts of terrestrial submarine groundwater discharge (SGD) were measured, but large material fluxes were seen by total SGD in some Asian coastal cities. The spatial variation of SGD was estimated around each city using a topographic model and radon measurements. Based on the accumulation and transport of pollutants, we evaluated the vulnerability risk for all cities. Pollution accumulation and transport were controlled by natural factors such as topography, climate, and geology as well as human impacts including pumping rate and pollution load. Core sampling in the coastal zone and groundwater sampling were done so as to reconstruct the history of contamination in each study area. Interpretations of chemical components and stable isotopes from the groundwater in Bangkok, Jakarta, and Manila revealed the origin of the groundwater and degree of nitrogen/ammonium contamination.

Analyses of temperature profiles measured in boreholes are also included as information on the history of past ground surface temperature (GST). Increased subsurface thermal storage showed the starting time and degree of surface warming due to the combined effects of global warming and urbanization. In the Bangkok area, the amount of GST increase is larger in the city center than in suburban and rural areas, reflecting the degree of urbanization. Subsurface heat storage, the amount of heat accumulated under the ground as a result of surface warming, is a useful indicator of the subsurface thermal environment. For example, we can compare the heat storage values measured at different times in the past with other parameters representing the urban subsurface environment that were obtained through other approaches.

Analyses by Geographical Information System of land cover/use changes due to urbanization are also included in this book. These analyses have been made for three different periods (1930s, 1970s, and 2000s) in Tokyo, Osaka, Seoul, Taipei, Bangkok, Jakarta, and Manila with a 0.5 km grid using nine different land cover/use types. Integrated indices of social changes, such as population and income (Driving force), groundwater pumping and dependency (Pressure), groundwater level (State), land subsidence (Impact), and regulation of pumping (Response), have been made on a yearly basis for seven cities over a period of 100 years (1900–2000). The development stages of a city are identified based on Tokyo using the DPSIR, and six other cities are compared with Tokyo for (1) land subsidence, (2) groundwater contamination, and (3) subsurface thermal anomaly. Groundwater storage and groundwater recharge rate data in seven cities have been compiled as integrated indices of natural capacities under climate and social changes. A stage model using the DPSIR framework revealed that Bangkok had the following benefit (relatively small damage with same driving force/pressure), Taipei had a higher natural capacity (higher groundwater recharge rate), and Jakarta had excessive development compared with Tokyo in terms of land subsidence.

By making comparisons of natural capacities (groundwater storage, groundwater recharge rates, etc.) under changing social and environmental indicators using the DPSIR model, we can see that Asian coastal areas have a positive potential in terms of groundwater use. It is thus possible to manage the groundwater resources in a sustainable fashion in this region.

I would like to acknowledge all members of RIHN who helped make the project successful. In particular for counterpart members including Backjin Lee (KRIHS, Korea), Chung-Ho Wang (Academia Sinica, Taiwan), Fernando Siringan (University of the Philippines, Philippines), Somkid Buapeng (Ministry of Natural Resources and Environment, Thailand), Gullaya Wattayakorn (Chulalongkorn University, Thailand), and Robert Delinom (LIPI, Indonesia). I also acknowledge for critical reviews of chapters by Osamu Shimmi, Akihiko Kondo, Kazuo Shibuya, Kazunari Nawa, Tomoyasu Fujii, Seiichiro Ioka, Youhei Uchida, Yasukuni Okubo, Shunsuke Managi, Ryo Fujikura, Hiroshi Matsuyama, and Kunihide Miyaoka. I also thank Ms. Yoko Horie of RIHN and Ms. Aiko Hiraguchi of Springer Japan for their help in editing this book. Some of the chapters have been translated in part from the book "Subsurface Environment in Asia" originally published in Japanese by Gakuho-sha (Makoto Taniguchi, editor). We are deeply grateful to the publisher for the permission granted for translation and for the copyright transfer of some of the figures.

Makoto Taniguchi
Research Institute for Humanity and Nature
Kyoto, Japan

Contents

Contributors

Yingjiu Bai (Chapter 12)
Tohoku University of Community Service and Science, 3-5-1 Iimoriyama,
Sakata, Yamagata 998-8580, Japan

William C. Burnett (Chapter 8)
Department of Earth, Ocean and Atmospheric Sciences, Florida State University,
Tallahassee, FL 32306, USA

Shu-Hao Chang (Chapter 10)
Institute of Earth Sciences, Academia Sinica, 128, Sec. 2, Academia Rd.,
Nangang, Taipei 11529, Taiwan, ROC

Supitcha Chanyotha (Chapter 8)
Department of Nuclear Technology, Faculty of Engineering, Chulalongkorn
University, Bangkok 10330, Thailand

Chieh-Hung Chen (Chapter 10)
Institute of Earth Sciences, Academia Sinica 128, Sec. 2, Academia Rd.,
Nangang, Taipei 11529, Taiwan, ROC

Deng-Lung Chen (Chapter 10)
Penghu Station, Central Weather Bureau,
2, Xinxing Rd., Magong, Penghu 88042, Taiwan, ROC

Robert M. Delinom (Chapter 6)
Research Center for Geotechnology, Indonesian Institute of Sciences, Cisitu
Campus, Jln. Cisitu – Sangkuriang, Bandung 40135, Indonesia

Surapol Dhammasarn (Chapter 7)
Department of Groundwater Resources, Ministry of Natural Resources
and Environment, Bangkok, Thailand

Takahiro Endo (Chapter 14)
Graduate School of Life and Environmental Sciences, University of Tsukuba,
1-1-1 Tennodai, Tsukuba, Ibaraki 305-8571, Japan

Yoichi Fukuda (Chapter 5)
Department of Geophysics, Kyoto University, Kitashirakawa Oiwake-cho,
Sakyo-ku, Kyoto 606-8502, Japan

Jehn-Yih Juang (Chapter 12)
National Taiwan University, No. 1, Sec. 4, Roosevelt Road,
Taipei 10617, Taiwan, ROC

Shinji Kaneko (Chapter 13)
Hiroshima University, 1-5-1 Kagamiyama, Higashi-Hiroshima,
Hiroshima 739-8529, Japan

Akihiko Kondoh (Chapter 12)
Chiba University, 1-33 Yayoi-cho, Inage-ku, Chiba 263-8522, Japan

Anirut Ladawadee (Chapter 7)
Department of Groundwater Resources, Ministry of Natural Resources
and Environment, Bangkok, Thailand

Jann-Yenq Liu (Chapter 10)
Institute of Space Science, National Central University, 300, Jhongda Rd.,
Jhongli, Taoyuan 32001, Taiwan, ROC
and
Center for Space and Remote Sensing Research, National Central University,
300, Jhongda Rd., Jhongli, Taoyuan 32001, Taiwan, ROC

Oranuj Lorphensri (Chapter 7)
Department of Groundwater Resources, Ministry of Natural Resources and
Environment, Bangkok, Thailand

Gayl D. Ness (Chapter 2)
Department of Sociology, University of Michigan, Ann Arbor, MI 48109, USA

Shin-ichi Onodera (Chapter 9)
Graduate School of Integrated Arts and Sciences, Hiroshima University,
1-7-1 Kagamiyama, Higashi-Hiroshima, Hiroshima 739-8521, Japan

Jun Shimada (Chapter 15)
Graduate School of Science and Technology, Kumamoto University,
2-39-1 Kurokami, Kumamoto 860-8555, Japan

Yang-Yi Sun (Chapter 10)
Institute of Space Science, National Central University, 300, Jhongda Rd.,
Jhongli, Taoyuan 32001, Taiwan, ROC

Makoto Taniguchi (Chapters 1 and 8)
Research Institute for Humanity and Nature, 457-4 Motoyama,
Kamigamo, Kita-ku, Kyoto 603-8047, Japan

Tomoyo Toyota (Chapter 13)
Japan International Cooperation Agency, 10-5 Ichigaya Honmuracho,
Shinjuku-ku, Tokyo 162-8433, Japan

Chung-Ho Wang (Chapter 10)
Institute of Earth Sciences, Academia Sinica, 128, Sec. 2, Academia Rd.,
Nangang, Taipei 11529, Taiwan, ROC

Makoto Yamano (Chapter 11)
Earthquake Research Institute, University of Tokyo, 1-1-1 Yayoi, Bunkyo-ku,
Tokyo 113-0032, Japan

Akio Yamashita (Chapter 4)
Graduate School of Life and Environmental Sciences, University of Tsukuba,
1-1-1 Tennodai, Tsukuba, Ibaraki 305-8572, Japan

Ta-Kang Yeh (Chapter 10)
Institute of Geomatics and Disaster Prevention Technology,
Ching Yun University, 229, Jianxing Rd., Jhongli, Taoyuan 32097,
Taiwan, ROC

Horng-Yuan Yen (Chapter 10)
Institute of Geophysics, National Central University, 300, Jhongda Rd.,
Jhongli, Taoyuan 32001, Taiwan, ROC

Akihisa Yoshikoshi (Chapter 3)
Department of Geography, Ritsumeikan University, 56-1 Tojiin Kitamachi,
Kita-ku, Kyoto 603-8577, Japan

Part I
Subsurface Environmental Problems
and Urban Development in Asia

Chapter 1
What are the Subsurface Environmental Problems?

Groundwater and Subsurface Environmental Assessments Under the Pressures of Climate Variability and Human Activities in Asia

Makoto Taniguchi

Abstract Subsurface environmental problems, such as land subsidence, groundwater contamination, and subsurface thermal anomalies, are important aspects of human life in the present and future but have not been evaluated as yet. Interactions between surface/subsurface and subsurface/coastal environments under the pressures of climate variability and human activities have been analysed for the cities of Tokyo, Osaka, Bangkok, Manila, Jakarta, Taipei, and Seoul, which are in different stages of urbanization. Analyses from satellite GRACE data showed that land water storage in Bangkok decreased since 2002. Groundwater tracers and 3D numerical simulations of groundwater showed that the groundwater flow system in the urban aquifer has been highly disturbed by pumping, causing a vertical downward flux in the urban area. Subsurface temperatures observed in the study cities illustrate the magnitude and timing of surface warming due to global warming and heat island effects. The amount of the increase in surface temperature was found to be larger in the city center than that in suburban and rural areas, reflecting the degree of urbanization. Contamination histories in each city have been reconstructed from sediment studies of nutrient and heavy metal contaminations. Analyses of land cover/use changes show that urbanization caused a reduction of groundwater recharge and an increase in thermal transfer into the subsurface environment. Two groups of integrated indicators: (1) natural capacities; and (2) changing society and environments, were used to analyse the relationships between the developmental stage of the city and the subsurface environment. Comparing Tokyo with each city shows that some cities have a benefit by developing later and/or benefit from a natural capacity such as higher groundwater recharge rate as higher input to aquifer. However, excessive development in Jakarta causes severe damage by land subsidence. Groundwater and subsurface environments should be investigated for their adaptation and resilience to changing environment conditions. In addition,

M. Taniguchi (✉)
Research Institute for Humanity and Nature, 457-4 Motoyama, Kamigamo,
Kita-ku, Kyoto 603-8047, Japan
e-mail: makoto@chikyu.ac.jp

M. Taniguchi (ed.), *Groundwater and Subsurface Environments: Human Impacts in Asian Coastal Cities*, DOI 10.1007/978-4-431-53904-9_1, © Springer 2011

subsurface environments should be treated together with surface and coastal environments for better management and sustainable use.

1.1 Introduction

Climate variability and increased demand for groundwater as a water resource due to increasing population has resulted in many subsurface environmental problems including land subsidence, groundwater contamination and subsurface thermal anomalies. These problems have occurred repeatedly in Asian major cities with a time lag depending on the developmental stage of urbanization (Taniguchi et al. 2009a). Thus, one may be able to assess future scenarios if we can evaluate the relationships between subsurface environmental problems and the development stage of the city. Although surface waters are relatively easy to evaluate, the changes in groundwater and other parts of the subsurface environment remains a difficult task.

While global warming is considered as a serious environmental issue above the ground or near the ground surface, subsurface temperatures are also affected by surface warming (Huang et al. 2000). In addition to global warming, the "heat island effect" due to urbanization creates subsurface thermal anomalies in many cities (Taniguchi and Uemura 2005; Taniguchi et al. 2007; Taniguchi et al. 2009b). The effect of heat islands on subsurface temperature is a global environmental issue because increased subsurface temperature alters soil water and groundwater systems chemically and microbiologically through geochemical and geobiological reactions that are temperature sensitive (Knorr et al. 2005).

Water and thermal energy are separated at the earth surface into the atmosphere and geosphere (subsurface) depending on land cover and use. Therefore analysis of changes in land cover/use can indicate the factors controlling subsurface environments. For instance, the change in land cover/use from open fields to houses and other structures results in decreases in groundwater recharge due to an increase in impermeable layers at the surface. This change also causes the increase in thermal energy into subsurface environment due to the change in heat balance at the earth's surface.

This study was intended to assess the effects of human activities on the urban subsurface environment, an important aspect of human life in the present and future but not yet evaluated. This is especially true in Asian coastal cities, where population numbers/densities and water demand have increased rapidly and uses of the subsurface environment have increased. The primary goal of this study was to evaluate the relationships between the developmental stage of cities and various subsurface environmental problems, including extreme subsidence, groundwater contamination, and subsurface thermal anomalies. We address here the question of sustainable use of groundwater and subsurface environments to provide for better future development and human well-being.

The subjects (Fig. 1.1) and research methods used in this investigation are as follows: (1) Urbanization; relationships between the developmental stages of cities and subsurface environmental problems were assessed by socio-economical analyses

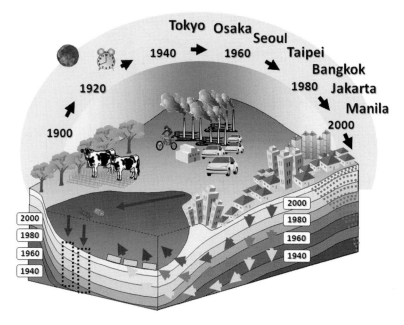

Fig. 1.1 Schematic diagram of the study theme

and reconstructions of urban areas based on GIS using historical records; (2) Groundwater; serious problems in subsurface environments and changes in reliable water resources were studied after evaluations of groundwater flow systems and changes in groundwater storage by use of hydrogeochemical data and in-situ/satellite-GRACE gravity data; (3) Material; accumulation of materials (contaminants) in the subsurface and their transport from land to ocean including groundwater pathways were evaluated by various techniques; and (4) Subsurface temperature; subsurface thermal contamination due to the "heat island" effect in urban areas was evaluated by reconstruction of surface temperature history and urban meteorological analyses.

1.2   Study Area and Methods

In order to assess the groundwater resources under the pressures of climate variation and human activities in Asia, intensive field observations on subsurface environments and data collections have been made in the urban basins including Tokyo, Osaka, Bangkok, Jakarta, Manila, Seoul, and Taipei (Taniguchi et al. 2009a, b, Fig. 1.2), where are mostly capitals and/or mega cities.

Subsurface temperatures have been measured in boreholes with one meter depth resolution to evaluate the climate change and heat island effects (Taniguchi and Uemura 2005; Taniguchi et al. 2007; Yamano et al. 2009). Temperatures were measured in 29 boreholes in Tokyo, 37 boreholes in Osaka, 15 boreholes in Seoul, and

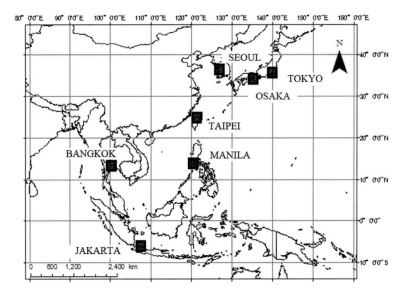

Fig. 1.2 Location of the study areas

14 boreholes in Bangkok. Thermistor thermometers, which can read temperature at a 0.01°C precision and an accuracy of 0.05°C, were used in these measurements. The diameter and depth of the boreholes in Tokyo are 8–20 cm (mostly 15 cm) and 126–450 m (mostly 200–300 m), respectively. Well diameters in Osaka are 10–40 cm (mostly 20 cm) and the depths range from 47 to 465 m (mostly 60–200 m). The depths of the boreholes in Bangkok and Seoul are 55–437 m and 498–968 m, respectively. The diameter of these boreholes is mostly around 10–30 cm. Therefore, no thermal free convection is expected (Taniguchi et al. 1999). The geology and other information concerning the aquifers in each city are given in Taniguchi et al. (2007).

One of the tools for evaluating the change in global/regional land water storage (LWS) is the use of satellite GRACE (Gravity Recovery And Climate Experiment) which was launched at 2002. Although GRACE is providing extremely high precision gravity field data from space, its spatial resolution is not enough to reveal the urban scale variations. However, recent study shows the GRACE data can be used for the study of water budget on a basin scale (Yamamoto et al. 2009).

To analyze the change of groundwater flow caused by the over-pumping in urban areas, a 3D numerical model with MODFLOW have been established in the area of Tokyo, Osaka, Bangkok, and Jakarta with the help of long term observation data of groundwater potential distribution. In addition to this, groundwater age tracers including CFCs and C-14 have been used to extract the time-series chemical records stored in the aquifer that has affected by the urban over-pumping resulting in accelerated groundwater flow.

Material contaminations in each aquifer are evaluated from the point of view of nutrient and heavy metal contaminations. Groundwater samplings in each city have been made for analyses of chemical components and stable isotope ratios, including N, C and Sr, to reveal the source of the contamination. Sediment cores near each city coast were sampled, and analyzed to reconstruct the contamination history back to approximately 1900.

In order to evaluate the surface environment which distributes water, heat, and materials into the subsurface, land use and cover conditions at three different periods (1930s, 1970s and 2000s) have been analyzed using GIS with a 0.5 km grid for each targeted cities. Land cover/uses were categorized into the following nine types: forest, house, industries, paddy field, other agriculture field, grass/waste land, ocean, water and wet land, and others.

To integrate all information concerning possible relationships between developmental stage of the city and subsurface environmental problems, we examined three subsurface environmental problems: land subsidence, groundwater contamination, and subsurface thermal anomalies. For each subsurface environmental problem, two sets of integrated indicators are used for the analysis. The first indicator is termed "natural capacities", that includes such characteristics as groundwater recharge rate and groundwater storage. The second indicator is referred to as "changing society and environment" and this includes population, income, industrial structure, groundwater dependency, groundwater pumping rate, groundwater level, and land subsidence.

1.3 Degradation of Groundwater Resources Due to Climate Variation and Human Activities

At the early stage of the urbanization, groundwater demand for industrial and domestic uses in the city increases because the groundwater is relatively inexpensive, has easy access, and provides stable and clean water resources. After increased groundwater consumption, the groundwater level decreases, and various subsurface environmental problems begins to occur such as land subsidence, groundwater salinization, dissolved oxygen reductions, and others. Thus, a shift in reliable water resources from groundwater to the surface water has occurred in many Asian cities (Fig. 1.3), due to an increase in demand of water resources (Taniguchi et al. 2009a). One of the reasons why these shifts occurred was due to severe land subsidence caused by excessive groundwater pumping in coastal Asian cities. Since the aquifers in these cities consist mainly of sedimentary formations, they are prone to subsidence. For example, the land subsidence in the Osaka plane has been observed since the 1930s due to excessive groundwater pumping for industrial use. The local government finally regulated this pumping after the 1960s. In the case of Bangkok, land subsidence was observed in the 1970s, but the government did not regulate pumping until the late 1990s.

On the other hand, climate change, such as changes in precipitation patterns due to global warming, have resulted in some Asian cities shifting their water resources in the opposite direction, i.e., from surface water to groundwater. For example, Taiwan is now using more groundwater because of the decrease in reliability of their surface water resources stored behind dams. In Taiwan, the decrease in number of precipitation days likely caused by global warming without changing the total amount of precipitation caused a decrease of reliability in their surface water

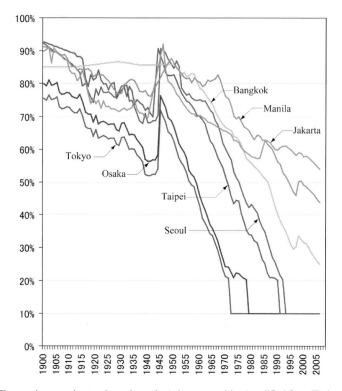

Fig. 1.3 Changes in groundwater dependency in Asian seven cities (modified from Taniguchi 2010)

resources. Therefore, at least in this case, climate change may be a principal reason for changing reliable water resources between surface water and groundwater.

To evaluate the regional groundwater resources, we analyzed updated GRACE data, not only for seasonal variations but also a secular trend of the mass variations in the Chao Phraya river basin (Bangkok). The result showed that the total mass change after 2002 was decreasing downstream of Chao Phraya and increasing upstream (Fig. 1.4). Although the GRACE trend agrees well with the global Terrestrial Water Storage (TWS) model (Fig. 1.4), there are some inconsistencies with regional or local land water models.

Groundwater tracers (CFCs and C-14) used in this study showed that the groundwater flow system in the urban aquifer has highly disturbed by the human pumping. A dominant vertical downward flux was revealed in the urban area by the CFCs and C-14, which originated from human activity in the urban area. Repeated measurement of C-14 of the groundwater in Jakarta show that the younger groundwater in shallower aquifer was directed downward due to groundwater pumping in urban area, making deeper groundwater younger (Fig. 1.5, Kagabu et al. 2010).

A 3D groundwater simulation (MODFLOW) showed a spatial change of the groundwater recharge area for Jakarta which was major recharge area of the pumped aquifer (Kagabu et al. 2010). This spatial change of the groundwater potential was

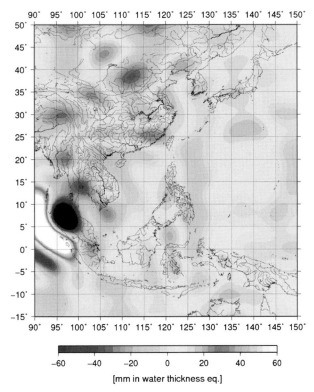

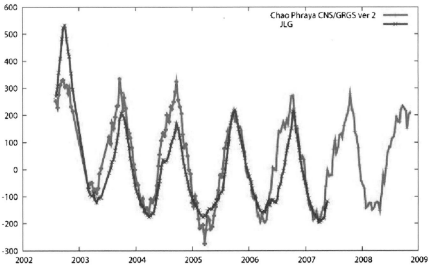

Fig. 1.4 Land water storage change by GRACE and comparison with TWS model (Yamamoto et al. 2009)

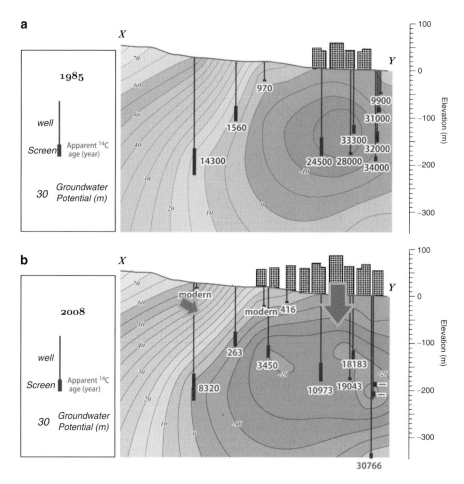

Fig. 1.5 Groundwater age obtained from carbon 14 at 1985 (**a**) and 2008 (**b**) (Kagabu et al. 2010)

strongly affected by the regional groundwater pumping regulations, and the success or failure of those regulations are mostly affected by the availability of alternative water resources for the city and the legal aspect of the groundwater resources for each nation.

1.4   Subsurface Thermal Anomalies Due to Surface Warming

Subsurface temperature is affected not only by global warming (Huang et al. 2000) but also by the heat island effect due to urbanization (Taniguchi and Uemura 2005). The combined effects of these two processes can have consequences on the groundwater system. Subsurface temperatures in four Asian cities (Tokyo, Osaka, Bangkok and Seoul) have been compared to evaluate the effects of surface warming due to urbanization and global warming, and the relationship to the developmental stage of each city (Taniguchi et al. 2007). Mean surface warming in each city ranged from 1.8°C to 2.8°C which was confirmed by air temperature monitoring over the last 100 years (Fig. 1.3). The depth of departure from the regional geothermal gradient was found to be deepest in Tokyo (140 m), followed by Osaka (80 m), Seoul (50 m), and Bangkok (50 m).

The increases in subsurface thermal storage in each city are shown in Fig. 1.6. As can be seen from these plots, the change in subsurface thermal storage was greatest in Tokyo, followed by Osaka, Seoul, and Bangkok. Numerical analyses using a one-dimensional heat conduction theory (Taniguchi et al. 2007) with different magnitudes and timings of the initiation of surface warming showed that the subsurface thermal storage was greater when the magnitude of surface warming was higher, and when the elapsed time from the start of surface warming due to urbanization was longer (Fig. 1.6). This trend was confirmed by air temperature records from each study area during the last 100 years (Taniguchi et al. 2007).

The heat island effect due to urbanization on subsurface temperatures is an important global groundwater quality issue, because it may alter groundwater systems geochemically and microbiologically (Knorr et al. 2005). Many cities in the world have this problem, particularly in Asia, where population has been increasing rapidly. Reconstructions of the surface warming history by use of

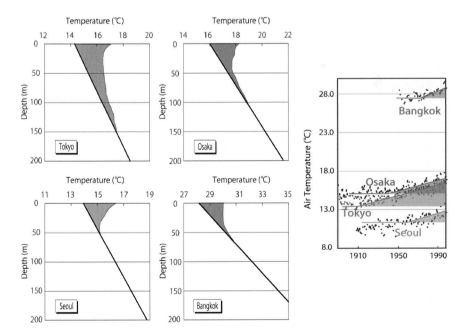

Fig. 1.6 Changes in air temperature and subsurface temperature in Bangkok, Seoul, Tokyo, and Osaka over the last 100 years

subsurface temperatures have been made in rural, suburban, and urban areas of Bangkok (Yamano et al. 2009). The results show that the surface warming started earlier in the current urban area, followed by current suburban area, and then the rural area. Therefore, a record of the expansion of the city can be preserved in these heat island records.

Subsurface heat storage, the amount of heat accumulated under the ground as a result of surface warming, is a useful indicator of the subsurface thermal environment and we can compare its values at specific times with those of other parameters representing urban subsurface environment obtained through various approaches.

1.5   Material Contamination

Nutrient and heavy metal contamination in each aquifer were analyzed and evaluated. Compositions of nitrogen contamination of the groundwater in each city show the nitrate contamination dominated in Jakarta, on the other hand ammonium contamination is dominant in Bangkok (Fig. 1.7, Umezawa et al. 2009).

Stable isotope ratios of N and C of groundwater can be used to indicate the origin of the contamination in Manila, Bangkok, and Jakarta (Umezawa et al. 2009) such as domestic or farm origin, and denitrification processes in Bangkok (Umezawa et al. 2009). Nitrite contamination was found in Jakarta and Manila, on the other hand, denitrification was found to occur in Bangkok even though huge loads of nitrogen. This is attributed to the natural conditions whether oxic or anoxic depending on the land slope (geomolophology), geology (volcanic or sediment), and hydroclimate codition (Hosono 2010).

The sediment cores from near the coast of the city were sampled and analyzed to reconstruct the contamination history back to approximately 1900 (Hosono 2010). Organic pollution and metal pollution histories were reconstructed in Asian cities, showing the changes in the C/N ratios and lead pollution depending on the development stage of the city. Some of the controlling factors included magnitude of loads, regulation of the loads, and others.

Groundwater salinizations were also found in Osaka, Bangkok and Jakarta. The difference of marine alluvium volume (same as topographic gradient), natural recharge and intensive pumping period controlled the degree of salinization. On the other hand, less terrestrial submarine groundwater discharge (SGD) but huge material flux by total SGD was found in Asian coastal cities (Burnett et al. 2007). Spatial variation in SGD was estimated around each city, using topographic models and Rn measurements.

Based on the accumulation and transport of pollutants, we evaluated the "vulnerability risk" in all cities. For example, relatively higher risks of nitrate are found in Jakarta and Manila, and arsenic pollutions were found in other cities, depending upon the redox state (Hosono 2010). The pollution accumulation and transport were controlled by natural factors such as topography, climate and geology as well as human impacts such as pumping rate and pollution load.

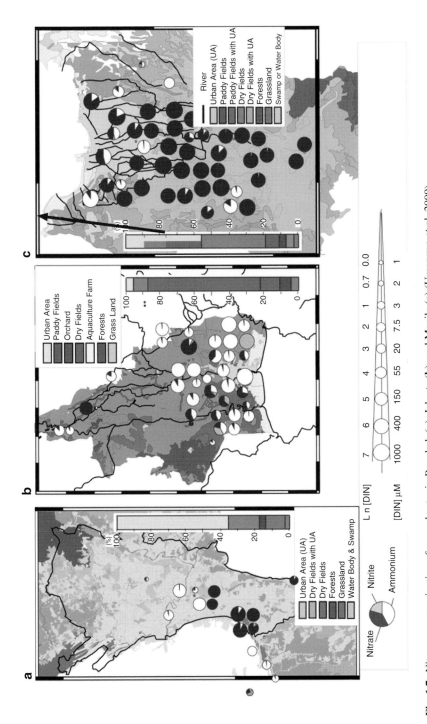

Fig. 1.7 Nitrogen contamination of groundwater in Bangkok (**a**), Jakarta (**b**), and Manila (**c**) (Umezawa et al. 2009)

1.6 Land Cover/Use Analysis for Subsurface Environments

The results of expanding urban areas in seven Asian cities, including Tokyo, Osaka, Seoul and Bangkok are shown in Fig. 1.8 and Table 1.1. As can be seen from these data, the urbanized areas expanded in Tokyo from 1930 to 1970 by 753 km², from 1970 to 2000 by 2,865 km², in Osaka from 1930 to 1970 by 569 km², from 1970 to 2000 by 907 km², and in Seoul from 1930 to 1970 by 196 km², from 1970 to 2000 by 954 km², respectively. Generally, the extension of urban areas decreases the permeable layers, thus reducing the groundwater recharge rate. The extension of urbanized regions also results in an increase of the thermal index which illustrates the magnitude of the heat island effect.

The analysis of land cover/use changes (urbanized area) shown in Table 1.1 indicates that the magnitude of area changed was larger during the period from 1970 to 2000 than during 1930 to 1970. Therefore, urbanization is accelerated more from 1970 to 2000 than from 1930 to 1970. The magnitude of the change in area is also greater in Seoul than in Osaka from 1970 to 2000.

The reliability of groundwater as a water resource may be decreased after urbanization due to reduction of the groundwater recharge rate, and increase of groundwater contamination including subsurface thermal anomalies. However, a new subsurface environmental problem has now occurred in Tokyo and Osaka, the "floating subway station" due to buoyancy by recovering groundwater levels after regulation of the pumping. This is attributed to the enough groundwater recharge to be recovered in the cities which are located in Monsoon Asia. Therefore, the development of integrated indicators are necessary for better understanding the relationship between human activities and the subsurface environment, and proper management.

1.7 Integrated Indicators

We developed two groups of integrated indicators to analyse the relationships between development stage of cities and subsurface environment: (1) changing society and environments, and (2) natural capacities. To analyze the relationships between development stage of the city (indicators of changing society and environment) and subsurface environmental problems, the DPSIR (Driving force, Pressure, State, Impact and Response) model is used for the seven targeted cities. The relationships between social economic stage such as population, and income, and subsurface environmental problems such as land subsidence, contamination and thermal anomalies, were framed for the analyses.

According to the degree of urbanization based on population, economy, groundwater dependency, subsurface problems such as land subsidence, and regulation/law, five development stages were categorized. The first stage is the early urbanization with higher groundwater dependency (more than 50 %), and the second stage is represented by increased water demand due to heavy industries. The third stage is the

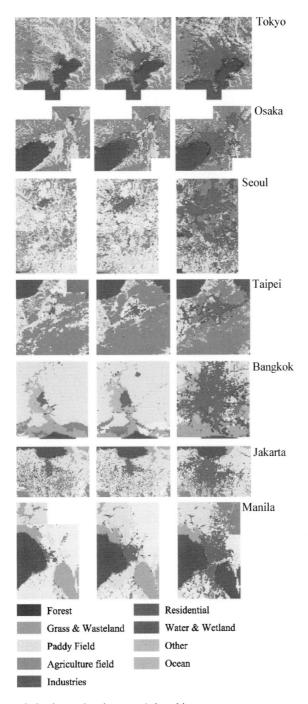

Tokyo

Osaka

Seoul

Taipei

Bangkok

Jakarta

Manila

■ Forest	■ Residential
■ Grass & Wasteland	■ Water & Wetland
■ Paddy Field	■ Other
■ Agriculture field	■ Ocean
■ Industries	

Fig. 1.8 Changes in land cover/use in seven Asian cities

Table 1.1 Changes in urbanized area including houses and industries in seven Asian cities

	1930		1970		2000		1930→1970	1970→2000
	(km²)	(%)	(km²)	(%)	(km²)	(%)	(km²)	(km²)
Tokyo	891.3	6.2	1,644.5	11.4	4,509.0	31.3	753	2,865
Osaka	321.0	4.7	859.3	12.6	1,716.5	25.2	538	857
Seoul	72.5	2.0	264.3	7.3	1,214.8	33.7	192	951
Taipei	47.8	2.0	73.3	3.1	228.5	9.5	26	155
Bangkok	41.3	1.7	79.3	3.3	930.3	38.9	38	851
Jakarta	303.0	5.1	377.0	6.3	1167.8	19.5	74	791
Manila	78.3	2.4	214.3	6.0	638.5	17.7	136	424

Percentage shows the ratio of urbanized area of total area

Table 1.2 Groundwater natural capacity in seven Asian cities

	Aquifer thickness (m)	Area (km²)	Storage (M ton)	Potential recharge rate (mm/year)	Residence time (year)
Tokyo	600	622	75	350	1,700
Osaka	1,200	222	53	430	2,800
Seoul	100	605	12	340	290
Taipei	100	272	5	640	160
Bangkok	500	1,569	157	450	670
Jakarta	100	740	15	640	160
Manila	300	632	38	820	370

The groundwater recharge rates were calculated with precipitation, evapotranspiration, and surface runoff from JRA

period when subsurface problems such as subsidence was first noticed and monitoring began. The fourth stage is the time when regulation of groundwater pumping began. The fifth and final stage is that period when groundwater is recovering and in some cases new subsurface problem (e.g., floating subway stations) started. According to these stages, Tokyo and Osaka are now in Stage 5, Seoul and Taipei are in Stage 4, Bangkok is in Stage 3, and Jakarta and Manila are in Stage 2.

In order to evaluate the natural capacities for the groundwater in each area, the groundwater storage (GS) with aquifer size and porosities, groundwater recharge (GR) rate which is the net result of evapotranspiration and surface runoff from precipitation, and the residence time (RT) which is calculated from groundwater storage and residence time (RT = GS/GR) were evaluated in the seven targeted cities, and the results are shown in Table 1.2. As can be seen, higher GS was found in Tokyo, Osaka and Bangkok, on the other hand, higher GR was found in Taipei and Manila followed by Bangkok and Jakarta.

Comparing Tokyo with each city shows that some cities benefit as developing later and from natural capacities such as higher groundwater recharge rate. However, the excessive development in Jakarta causes severe damage from land subsidence. Groundwater and subsurface environments as alternative, adaptation, and resilience to the changing environment should be treated with surface and coastal environments for better management and sustainable use.

Comparisons with natural capacities (groundwater storage, groundwater recharge rate, etc.) and changing society and environment indicators depending on the DPSIR model for urban groundwater problems in Asian cities clearly demonstrate that Asian coastal areas have the good potential for groundwater recharge, and it is possible to manage the groundwater resources sustainably in this region.

1.8   Conclusion

Several new methods have been developed in this study to evaluate the sustainable use of groundwater in Asia. Satellite GRACE, groundwater tracers, and 3D numerical simulations revealed the regional current groundwater status of the targeted areas in Asia, and showed the induced downward groundwater flow due to excessive pumping Global warming and heat island effects as human and climate impacts on subsurface environments have been evaluated in several Asian cities. The analysis of subsurface temperatures showed that the subsurface thermal storage was greater when the magnitude of surface warming is higher and the elapsed time from the start of surface warming due to urbanization was longer. Analysis of land cover/use changes in seven Asian cities (Tokyo, Osaka, Seoul, Bangkok, Taipei, Jakarta and Manila) shows that urbanized areas expanded much faster from 1970 to 2000 compared to the measured increases from 1930 to 1970. Urbanization causes a decrease in the groundwater recharge rate and increases thermal transport into the subsurface environment. In order to develop integrated indicators for better understanding the relationships between human activities and the subsurface environment, the DPSIR model is used to analyze the relationships between social economic and subsurface environments depending on the stage in the model. Comparisons with natural capacities and changing society and environment indicators based on the DPSIR model for urban groundwater problems in Asian cities clearly demonstrate that Asian coastal areas have good potential for groundwater recharge, and it is possible to manage the groundwater resources sustainably in this region .

Acknowledgements   The author thanks the members of the USE (Urban Subsurface Environment) project of RIHN (Research Institute for Humanity and Nature) for helping to conduct this research. This investigation is closely connected with other international research programs, including UNESCO-GRAPHIC (Groundwater Resources under the Pressures of Humanity and Climate Changes).

References

Burnett WC, Wattayakorn G, Taniguchi M, Dulaiova H, Sojisuporn P, Rungsupa S, Ishitobi T (2007) Groundwater-derived nutrient inputs to the Upper Gulf of Thailand. Cont Shelf Res 27:176–190
Hosono T (2010) The nitrate-arsenic boundary as an important concept in aquatic environmental studies. In: Taniguchi M, Shiraiwa T (eds) The dilemma of the boundary. Springer (in submission)

Huang S, Pollack HN, Shen Po-Yu (2000) Temperature trends over the past five centuries reconstructed from borehole temperatures. Nature 403:756–758

Kagabu M, Shimada J, Nakamura T, Delinom R, Taniguchi M (2010) The groundwater age rejuvenation caused by the excessive groundwater pumping in Jakarta area, Indonesia. J Hydrol (under submission)

Knorr W, Prentice IC, House JI, Holland EA (2005) Long-term sensitivity of soil carbon turnover to warming. Nature 433:298–301

Taniguchi M, Shimada J, Tanaka T, Kayane I, Sakura Y, Shimano Y, Depaah-Siakwan S, Kawashima S (1999) Disturbances of temperature-depth profiles due to surface climate-change and subsurface water flow; (1) An effect of linear increase in surface temperature caused by global warming and urbanization in Tokyo metropolitan area, Japan. Water Resources Research 35:1507–1517

Taniguchi M, Uemura T (2005) Effects of urbanization and groundwater flow on the subsurface temperature in Osaka, Japan. Phys Earth Planet Inter 152:305–313

Taniguchi M, Uemura T, Jago-on K (2007) Combined effects of urbanization and global warming on subsurface temperature in four Asian cities. Vadose Zone J 6:591–596

Taniguchi M, Burnett WC, Ness GD (2009a) Integrated research on subsurface environments in Asian urban areas. Sci Total Environ 404:377–392. doi:10.1016/i.scitotenv.2009.02.002

Taniguchi M, Shimada J, Fukuda Y, Yamano M, Onodera S, Kaneko S, Yoshikoshi A (2009b) Anthropogenic effects on the subsurface thermal and groundwater environments in Osaka, Japan and Bangkok, Thailand. Sci Total Environ 407:3153–3164. doi:10.1016/j.scitotenv.2008.06.064

Taniguchi M (2010) Subsurface environmental problems. In M. Taniguchi ed., Subsurface environments in Asia. Gakuho-sha (Japanese)

Umezawa Y, Hosono T, Onodera S, Siringan F, Buapeng S, Delinom R, Jago-on KA, Yoshimizu C, Tayasu I, Nagata T, Taniguchi M (2009) The characteristics of nitrate contamination in groundwater at developing Asian-Mega cities, estimated by nitrate $d^{15}N$ and $d^{18}O$ values. Sci Total Environ 407:3219–3231

Wang C-H (2005) Subsurface environmental changes in Taipei, Taiwan: current status. In: Proceedings of RIHN international symposium on human impacts on urban subsurface environments, October 18–20, 2005, Kyoto, Japan, pp 55–59

Yamamoto K, Fukuda Y, Nakaegawa T, Nishijima J (2009) Landwater variation in four major river basins of the Indochina peninsula as revealed by GRACE. Earth Planets Space 59:193–200

Yamano M, Goto S, Miyakoshi A, Hamamoto H, Lubis RF, Monyrath V, Kamioka S, Huang S, Taniguchi M (2009) Study of the thermal environment evolution in urban areas based on underground temperature distributions. Sci Total Environ 407(9):3120–3128

Chapter 2
Asian Urbanization and Its Environments

Gayl D. Ness

Abstract RIHN's Human Impacts on Urban Subsurface Environments projects has made substantial progress. It is especially effective in seeing the unseen and gaining a better understanding of that unseen environment. The Asian cities on which the project focuses represent one of the most dramatic human social movements in our time. In the span of a mere century the Asian urban population will expand from 234 million to 3.2 billion. Moreover that urban population lives in a high risk environment, whose vulnerability increases with global warming. As many have observed, the struggle for sustainability will be won or lost in the world's urban areas, and the Asian urban environment will play a highly critical role in this struggle. We suggest that the RHIN project's excellent work on the unseen environment now needs to make links to the many other urban environments to promote the development of sustainable cities. Here we suggest two following steps. One is the development of a standard protocol that governments can use to assess the condition of the urban subsurface environment. The second is modeling exercises that will focus attention on linking the many urban environments to work toward sustain able cities.

2.1 Introduction

In this paper I wish to do two things. First, I wish to extol the excellent work the RIHN Subsurface Environment Project has done in addressing an unseen part of the Asian urban environment: its subsurface environment (Taniguchi et al. 2009a, b). This is work that must be recognized as one of those major achievements of scientists: to see the unseen.[1] The work is highly innovative, technically advanced,

---

[1] A fellow graduate student many years ago made the same point of Freud's great discoveries: the unconscious. How does one discover that of which we are not conscious!?

G.D. Ness (✉)
Department of Sociology, University of Michigan, Ann Arbor, MI 48109, USA
e-mail: gaylness@umich.edu

M. Taniguchi (ed.), *Groundwater and Subsurface Environments: Human Impacts in Asian Coastal Cities*, DOI 10.1007/978-4-431-53904-9_2, © Springer 2011

and especially enlightening. To see subsurface water levels rise and fall with urban processes and policies, to see underground pollutants rise and spread to ocean fringes, to recognize that rapidly growing cities exert a massive pressure on the land on which they sit, must rank as major discoveries today. I extol these superb advances.

But I also wish to ask that these advances be taken further; to do more with them; and especially to link these studies to the larger world of modern cities and their complex and increasingly expanding environments. And even more, I wish to ask that these studies be linked to the ultimate question we now ask: how to build sustainable cities and societies in our new global environment.

The cities this initiative has investigated are all Asian cities. I noted earlier (Taniguchi et al. 2009a) that Asian cities have a long and very distinctive history. Asia has been more advanced than other regions in urbanization for the past three millennia. It was only overtaken by the modern Western urban-industrial transition in the nineteenth century. That transition, and even its antecedents, saw western cities located on waterways and the sea, pointing outward for trade. Historically, Asian cities have been primarily inland cities. They represented strong political centers with considerable capacity to administer a large and productive hinterland that supported the cities.

I have also shown that this long history has produced a distinctive political culture, whose impact we can now see in the previous study. In the Subsurface Environments Project we have seen that Manila (Jago-on et al. 2009), and Cebu (Flieger 2000), lack the political capacity to control private well digging. This is a capacity found in all the other cities of the RHIN studies. That difference in political capacities can be traced to the two millennia history of Asian civilization and urbanization. Indians brought ideas of kingship to all Southeast Asia nearly two millennia ago. This helped produce the centralized political systems we have seen throughout the region for the long past.[2] But Indians did not get to the Philippines, which, alone in Southeast Asia, lacks a history of strong, centralized political systems. The past is always with us; never fully determining our lives, but always affecting them.

Here I shall first review the larger scene of Asian urbanization, pointing out its distinctive history, and its present and massively distinctive current and near future conditions. This identifies the urban scene as the critical area in which the quest for sustainability will be won or lost. Then I shall examine a number of urban environments of the cities we have studied, to place the newly discovered subsurface environments in a larger, and highly related, context. I shall then review the larger political organization that leads us to examine environments both scientifically and with a view to managing them. This will take us into a range of political organizations by which humans view and attempt to manage their environments. Finally, I shall propose some next steps. How this wonderful examination of subsurface environments leads to a fuller, richer, and more useful capacity to manage the many urban environments.

---

[2] See especially in this context, Lieberman (2003).

2.2 Asian Urban History

The Subsurface Environments project has been examining seven *mega* cities (over five million population): Bangkok, Jakarta, Manila, Osaka, Seoul, Taipei and Tokyo. These seven mega cities are a part of a distinctive population dynamic in Asia that has emerged only in the past half century. What we are seeing is an extremely rapid increase of large cities and urban populations. Both the speed and the magnitude of the Asian urban growth are remarkable.

In 1950 there were just two mega cities in Asia (Tokyo and Shanghai), and only eight in the entire world (add New York, London, Paris, Moscow, Buenos Aires and Chicago) (UN 2005). The combined population of these mega cities was 59 million for the world and 17 million for the Asian cities. By 2000 there were 17 mega cities in Asia and 42 in the world as a whole. The combined population of all mega cities was 408 million, up 600% in just 50 years. The combined population of the Asian mega cities was 195 million, an increase of over 1,000% in half a century.

The overall urban population continues to grow worldwide and to grow more rapidly in Asia. Not only the speed, but the magnitude is impressive. It took Europe 250 years to become urbanized; its urban population grew from a mere seven million in 1750 (Chandler and Fox 1974) to 281 million in 1950, when the region was 52% urbanized. In the next half century it completed the urbanization process as its urban population grew to 520 million, and urbanization reached over 80% of the population. By contrast all Asia had only 234 million urbanites in 1950, making it 16% urbanized. By 2000 the Asian urban population had grown six times to 1.2 billion, making the region 36% urbanized. In the next half century we can expect the urban population to more than double to 3.2 billion, making the region 65% urban. In sum, Europe took 300 years to get to half a billion urbanites; Asia will gain three billion urbanites in only a century!

All urban areas continue to grow throughout Asia. Even with the significant slowing of overall population growth,[3] urban areas continue to grow. The urban population will continue to grow even as the population in rural areas is actually shrinking. Already in 2000 five of the 14 major Asian countries showed negative rural population growth: Indonesia, Philippines, China, Japan and South Korea (Ness and Talwar 2004). By 2030 seven more will be added to this list: Bangladesh, India, Sri Lanka, Malaysia, Myanmar, Thailand, and Vietnam. Even while rural populations decline in absolute numbers, the urban population will continue to increase.

Urbanization is the great reality of our times. Asian urbanization will continue even when overall populations begin to decline. And the numbers will be massive: from 234 hundred million in 1950 to a projected 3.2 billion in 2050.

---

[3] The overall Asian population growth rates peaked at 2.25% in 1970–1975, when the population was 2.1 billion; the overall population will continue to grow through 2030. The East Asia population growth rates slowed from their peak of 2.26% in 1970 to −0.40 in 2000. Its population is projected to begin to decline after reaching 1.663 billion in 2030.

But of course it is not merely a matter of numbers of people. As I wrote in our last engagement, the capacity to build on the urban surface has also mushroomed in the past century. The shift from stone or masonry walls to steel skeletons has given us a great capacity to build upwards. Masonry walls can support heights of 167 m. The Mondanook Building in Chicago built 1901 is the tallest; Chartres Cathedral, from the sixteenth century soared 113 m. After 1900 the steel skeletons could support 200 m and more. Now 400 m is not the limit; Taipei's Taipei 101 rises to 501 m. Again the current speed of development is the real news. Eighty-eight of the world's 200 tallest buildings have been built only since 2000.

In addition there is the physical capacity pump water to serve the needs of the city. As we have seen earlier, this has produced substantial subsidence, which has been curtailed and sometimes reversed when the political system has sufficient capacity to curtail private well digging. Add now the chemical and production advances that have made pollution one of the major problems of cities today. This is a problem for the subsurface environment, well documented by the previous RIHN publication.

If this were not enough, climate change poses severe threats to all, but especially to the many Asian urban areas in low lying lands. Climate change inevitably means increasing temperatures, increasing variability in weather patterns with increases in extreme events, and rising sea levels. Though historically Asia's major cities were primarily inland cities, since the nineteenth century most of the rapid growth has been on the seaside, and along major rivers leading to the sea. Studies of areas at risk to climate related hazards show especially high concentrations in Asia: Japan, Eastern China, all Southeast Asia, and India and Bangladesh on the Bay of Bengal (UNFPA 2007). Moreover, large and growing proportions of people and cities are situated in Low Elevation Coastal Zones (LECZ), where the threats of climate change and sea level rise will be most pronounced. These are also areas in which the subsurface environment will be especially important. In Asia some 15% of all people and 18% of urban people live in such vulnerable zones. Three percent of all Asian lands but 12% of urban lands are in these vulnerable zones. These are the largest percentages of all world regions except for Australia and the small island states of the Pacific and Indian Oceans.

Observations such as these led Maurice Strong to proclaim:

"The struggle for sustainable development will be won or lost in the cities of the world"[4]

Worldwatch made a similar observation: "It is particularly ironic that the battle to save the world's remaining healthy ecosystems will be won or lost not in the tropical forests or coral reefs that are threatened, but in the streets of the most unnatural landscapes on the planet," (Worldwatch Institute 2007, p xxiv).

The basic message here is that the cities of the RIHN project constitute a highly critical area for the future of sustainability. If the future of the struggle for

---

[4]UNCED 1992.

sustainability will be won or lost in the cities of the world, a great deal of that struggle will take place in Asia, among cities like those the Subsurface Environment project has been studying.

2.3 Urban Environments and Their Social Organizations

But to use these studies to promote the development of sustainable cities, they must be linked to studies of other environments and to sustained *organized* efforts to manage the human impacts on the broader urban environment. Those other environments include the air, land and water, precipitation, the natural landscape and the built up environment. The built up environment is itself a highly diverse blend of structures and functions: factories that produce goods, residential units, structures for a wide variety of human services, and a large and complex transportation system both subsurface and above ground and in the air. It includes a natural environment that is increasingly controlled by human activity: dams for reservoirs, built up river banks to control water flow, sea sides built for port facilities and recreation.

These above surface environments are all linked to the subsurface environments in highly complex ways. More importantly, many of these different environments have come under the control of local or national, or even international political systems. In some cases, those political systems have greatly advanced environmental welfare through various forms of protection. In other cases, equally revealing, the political system seems incapable of protective action. Many have been suggested in the past RIHN studies. Let me just list a few. I will examine some of our past studies and ask what other parts of the urban environment are linked to the subsurface environment. This will only serve to illustrate the great complexity of the many urban environments that interact with the subsurface.

It was from Karen Jago-on et al. (2009) that I found Manila has no capacity to control private well digging. This paralleled a study I was involved in a decade ago (Flieger 2000) where we found the same thing in Cebu, the Philippines' second major sea port. Here is a political system that has been called *An Anarchy of Families* (McCoy 2009). The kinship system essentially dominates the political system. Everything is for the family, nothing for the larger community: province or nation. We can trace this system to the failure of Indian civilization to get to the Philippines with its ideas of kingship near two millennia ago. It can also be traced to the legacy of Spanish and American colonialism. From the other Asian cities in this study, we find something diametrically opposed to Manila: these cities have the political capacity to act for the larger community. How do we link this view of political systems more directly to the problems of the subsurface environment?

Yoshikoshi et al. find Bangkok less affected in hydro environmental changes than Tokyo or Osaka. This is largely due Bangkok's later urban development, which means less expansion of impermeable surfaces. By now, their study shows, virtually all Tokyo and Osaka land is covered with impermeable surfaces, compared with about 40% in Bangkok. Tokyo and Osaka are much older cities. Bangkok's development is much more recent. I recall walking the streets of Bangkok in the 1960s when many klongs were being covered over to make wide thoroughfares. Bangkok's development is late and exceedingly rapid. It has a highly congested urban road system that is also being expanded very rapidly. It has lagged in the development of rapid mass transit systems. Only in the past few years has it developed a more extensive system of elevated highways and electric trains. Its historical position and its current political administrative system have given us less impermeable surface, but it is unclear how long this will last. At what rate is this developing? What will be the future impacts?

There is another experience that relates to the issue of impermeable surface areas. Chennai, India has developed an extensive recharge system that funnels rainfall from virtually all of the city's impermeable surfaces into the subsurface environment. All impermeable surfaces, from roof of homes and, buildings, to roadways and parking lots are drained into 3-m deep rock-filled wells that use the runoff to recharge the underground water table. Since this system has been in place subsurface water levels have risen substantially. How can we link subsurface environments with the great variety of possibilities for managing land surfaces?

There is yet another issue on Bangkok that brings local climate into the question. Yoshikoshi et al. also find Bangkok has more abundant rainfall than Tokyo or Osaka. But of course that rainfall is not evenly disturbed throughout the year. It is a monsoonal rain region with heavy precipitation during the summer, southwest, monsoon and little during the winter, northeast, monsoon. The summer monsoon rains can mean 20 or more days of rain a month, bringing easily 200 mm of rain. The dry winter months may see 1 or 2 days of rain, with no more than a few millimeters. These heavy rains do two things. They clean the air of suspended particulate matter, nitrous oxides and ozone. Then they leach these substances through the soil, carrying with them any other soluble materials in the soils through whatever drainage plumes are found in any particular place. They may well show up in the seepage that Burnett et al. find in Bangkok's subsurface and canal flows. They may enter the river and flow away. For much of Asia monsoons produce highly varied seasonal rainfall. What is the impact of both subsurface and terranean flows of water and pollutants?

Warming. A number of the studies deal with warming: temperature, urban heat islands, historical trends, subsurface and surface variations and so on (Yoshikoshi et al. 2009; Kataoka et al. 2009; Yamano et al. 2009; Huang et al. 2009; Taniguchi et al. 2009a, b). We find nearly universal trends in rising urban temperatures. Urban centers are warmer than their suburbs. Urban diurnal variations decline over time. Some of this is distinctly urban, what goes on in large cities; some is more directly related to long term warming and climate change. Here is a major problem that calls for large scale, interdisciplinary study. Where does urban heat come from? What different urban environments produce heat? Transportation provides some but in many cities it is the buildings themselves that produce the most heat. There are countless electric motors for elevators, escalators, air conditioning and heating and lighting. What public policies, from utility rates to formal regulations, to sources of electricity (from hydro to natural gas to dirty coal) affect the amount of heat generated for unit of work? What technological developments, from what sources, produce the motors that use electricity? China has recently unveiled a super-green building where the elevators sense the weight in them and direct the motors to use the appropriate amount of energy to lift or lower the load. It should not be difficult to calculate how much energy such a system will save. Elevators are all subject to government control, licensing and inspection. This means that somewhere in the urban administrative system there are records of all elevators and how many flights they traverse. A sample of motors that drive the elevators could provide us with a gross estimate of energy use per elevator, from which we could easily calculate savings in energy use from the new systems. This is, of course, only one set of questions that focus on the source of heat. Another set of questions must focus on the consequences of the heat. Here human health is a major concern, especially as cities face major heat waves in the summer, which have been known to kill directly.

Lee et al. (2009) examined nutrient flows from both subsurface and surface water into Masan Bay in South Korea. What was striking was the finding that subsurface flows equaled or greatly surpassed the flows from surface water. (Burnett found similar significant pollution flow through subsurface conduits in Bangkok.) Here were nutrients that caused massive red tides eutrophication and algae blooms. Government efforts to control the runoff have yet to be successful. Perhaps it is because they have failed to understand the importance of the subsurface flows. This provides us with another very important conduit for pollution flows. The next questions lead us to ask about the sources of the nutrient flows, and why government regulations have been unable to stem the tide. But they also raise another question at the end of the flows. What are the health and economic consequences of the red tides and eutrophication? The discovery of subsurface nutrient flows provides important information on a long chain of events, and the entire chain must be researched.

Much more could be said about the many environments of the urban scene and of their social organization, but I think this is sufficient to suggest that those other environments are an important element of the subsurface environment. How can these many environments be linked together more effectively? For in a sense they must be linked together if we are to have any impact in developing more sustainable cities and societies. Let me first review briefly the technological and social developments that have come so massively and rapidly in less than the past half century. Then I shall make a modest two part proposal.

2.4 Technological and Social Developments[5]

The past century has seen dramatic developments in the technology and the social organization that affect what we are now calling sustainability. These developments are not just dramatic. They are mind boggling. Let me note just

---

[5] By Social organization I mean the broader political, social economic and cultural organization of the human species.

two of the major technological development of the past century, then talk about recent social developments.

A century ago we were just being introduced to the automobile and its wonderful internal combustion engine. It is difficult to envisage now, but remember that at that time the automobile was hailed as a wonderfully clean and safe form of transportation. Compared to the horse, this was certainly true. Think of what Tokyo would look like now if it had the same ratio of horses to people that it had in 1900. Tokyo Bay would be totally horse urine. The city's major problem would be the removal of thousands of tons horse manure, whose immense piles would attract swarms of flies that would make gastrointestinal diseases pandemic. Today's cars are much cleaner, to be sure. Brain damaging lead has been removed and all manner of exhaust pollution (NOX, SPM, etc.) has been filtered out. The internal combustion engine is now known to be a wonderfully powerful source of an efficient and effective global transportation system. It is also known as a major source of the green house gasses that are giving us an unprecedented human impact in climate change.

Next consider Chlorofluorocarbons. In the 1930s when they were discovered and used in air conditioning, they were considered a wonderful product. They were odorless, non-corroding, non toxic, and they apparently simply went away. And they gave us refrigeration and air conditioning that have contributed massively to our global development. Would Bangkok, Manila, Surabaya, Osaka, Tokyo, Taipei or Seoul be the cities they are today without air conditioning and refrigeration? Then in the 1970s we found, first by theory then by measurement that the wonderful CFCs did not simply go away. They went to the stratosphere and destroyed ozone, that shield that keeps the earth safe from the sun's ultraviolet rays. That discovery led to a major global political realignment, in which the USA led other nations to the ban on CFCs.[6]

Next consider the dramatic development of political organizations concerned with *environmental protection*. In 1962 Rachael Carson wrote, *Silent Spring*, which is often credited with awakening the world to the dangers of chemical pollution. Eight years later the USA created the Environmental Protection Agency. In 1971 the Japanese government established an Environmental Agency, which was upgraded to Ministerial level in 2001. In 1972 the United Nations convened an international conference on the environment, which led to the creation of the United Nations Environment Program (UNEP). Three years later the USA and Japan signed an agreement to cooperate on environmental protection. Since 1970 there has been a very rapid development of scientific research organizations under the auspices of national governments and universities. Since then virtually every country has developed some version of an environmental protection agency. Non-governmental organizations proliferate throughout the world. The growth of organizations in the field of environmental protection is like a mushroom cloud, though hardly as deadly.

---

[6] The issue is highly complex and the US leadership in the CFC ban was not an altruistic act. In fact, DuPont had developed an effective alternative and if the ban were in effect, it stood to gain a very large market share of the substitute for CFCs.

RIHN is itself one of the manifestations of this trend. The world has mobilized massive resources to study our environments and the human impacts on them. The subsurface environment project reflects clearly the strong focus on not just environmental changes, but more specifically on the human impacts of those environmental changes.

There are next steps, however, of which I should like to suggest two, based on the RHIN work. First, I suggest we develop a standard measurement or assessment protocol that could be used by all or many cities to assess the health of their subsurface environment.

Second, there are scores of environmental changes that can be assessed in terms of the human activities that drive those changes. But many lead to the larger question of a sustainable society or sustainable development. We are all now asking how we can leave to future generations a life as good as or better than the one we have now. Sustainable societies, sustainable cities, are the next items on the agenda. For this the procedure of modeling cities can be especially useful.

2.5 Next Steps

2.5.1 Protocol for Assessing the Health of the Subsurface Urban Environment

The Subsurface Environment Project has developed some excellent tools and procedures for assessing the condition of the subsurface environment. It might be useful to pause a moment and try to gather together these steps to develop a somewhat standardized protocol that could be proposed to and used by many cities to assess the health of their subsurface environment, What questions should be asked? What specific conditions should be measured? What are the best procedures or instruments for these measurements? What are the dangers or threats to the subsurface environment from typical urban development? How can these dangers best be assessed? And most important how can these dangers be countered by effective policies (for example, restricting ground water extraction)? These are some of the questions a standardized protocol would address.

This type of protocol development, based on RHIN findings, should be very useful to international organizations like UNEP, UNDP, The World Bank and the Regional Development Banks, and many of the bilateral aid programs.

2.5.2 Modeling Urban Sustainability

RIHN has already demonstrated great capacities in developing interdisciplinary studies of Humanity and Nature. It has also shown a great deal of creativity in developing powerful and useful research projects. Let me suggest one more, a modeling exercise aimed at promoting sustainable cities.

Modeling, or dynamic modeling, is a well known and usually highly productive intellectual exercise.[7] Think of the massive biogeochemical models that give us the climate models on which so much of modern environmental activity is based. Models are also commonly used in making economic, demographic, and health projections. Modeling also raises a number of critical and very useful questions. The Subsurface Environment Project has provided a base, both literally and figuratively for building a larger model of sustainable cities. Models are, to be sure, simplified constructs made to represent a specific set of processes. Simplification is the great strength of modeling, since it is literally impossible to think of all the conditions and processes that operate in any living system. Simplification allows to select a few conditions and processes and to examine their interactions. If we get the conditions and processes right, and if we understand the connections among them, we can produce future possible scenarios that will show us possible effects of current policies, processes and conditions. In effect, if we are concerned with *sustainability*, modeling is a clear necessity, because we must find a way to project past and current conditions and processes into the future. That is what modeling does.

There is another highly important advantage of modeling. In effect we all carry models in or heads. These are idea systems we have of how things work, what conditions or events produce the conditions we see. We have ideas, or personal models, of how economic development is promoted. We have similar models of what produces crime, illnesses, natural and manmade disasters, and even things like personal happiness. The problem with these personal models is that their assumptions are rarely made explicit and tested. It is precisely this that makes dynamic modeling so useful and powerful. Its assumptions are clearly stated and quantified. They can be examined and tested to refine their accuracy in ways that or personal models never achieve.

To illustrate, let me lay out a model of urban systems developed earlier for a quite different purpose (Ness and Low 2000).[8] This was developed for medium-sized cities to help their administrators recognize their current conditions and the future implications of those conditions. The model was first described as a closed system model, and later it was opened to illustrate what an open system might look like. The models were designed to help urban administrators in nine Asian cities[9] think about the future implications of current conditions.

We called this model a "metabolic model" of the urban system. By that we meant a system that uses resources to produce life. That is, the output of any

---

[7] I have produced a brief statement of modeling in Ness and Low (2000).

[8] It would be most interesting to consider similarities and differences with Kaneko's DPSIR modeling device, though that is beyond the scope of this paper.

[9] The Asian Information Center of Kobe (AUICK) in cooperation with UNFPA provides assistance to nine cities, known as AUICK Associate Cities. They are, from west to east, Faisalabad, Pakistan; Chennai, India; Chittagong, Bangladesh; Kuantan, Malaysia; Khon Kaen, Thailand; Danang, Vietnam; Surabaya, Indonesia; Olongapo City, The Philippines, and Weihai, China.

metabolic systems is life. This may be conceived as a dichotomous attribute: life or death. Or the outcome may be conceived as a variable: greater or lesser health, or something we can call The Quality of Life. Figures 2.1 and 2.2 below show the model first as a closed system and then as an open system model.

Figure 2.1 shows a closed system. The broad social environment is illustrated by the border; it is the Social, Political, Economic, and Cultural system of the city. This is a cumbersome title, but it is meant to reflect the kinds of differences we have

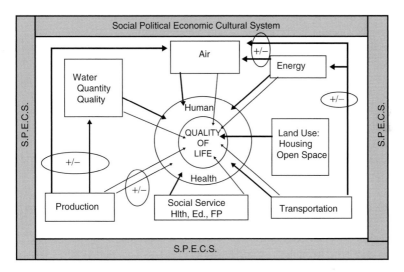

Fig. 2.1 A "metabolic model" of the urban system (closed system)

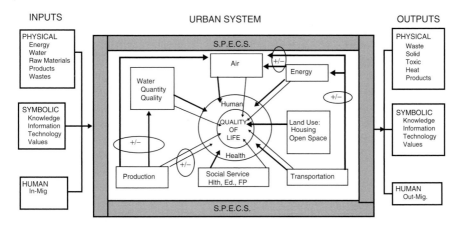

Fig. 2.2 An open systems model

seen in Manila's inability to stop private well digging (Jago-on et al. 2009), and the strength of that capacity in other cities. There is as yet no suitable taxonomy of these social systems that speaks to differences in their capacity to promote sustainable development. That is something that needs to be developed. I have some suggested ideas below. Inside the border of the current model are seven urban subsystems that affect the quality of life. The primary output of this system is the central circle, Quality of Life. There is an extensive literature on QOL, but we have yet to find a single variable measure that be used for many different environments. The next circle represents a much more measureable variable: human health. One could use a single variable measure, such as the Infant Mortality Rate, often considered the single best indicators of a society's health. One could also develop a composite, similar to the UNDP's Human Development Index.

The seven urban systems identified here were developed from extensive surveys of urban administrators, who indicated these were critical parts of the agencies they had to work with. These subsystems are connected by heavy lines, indicating direct effects, and lighter lines indicating indirect impacts. All lines have positive or negative valences. Where no valences are given, and in the link between water and the quality of life, these are assumed to be positive.

As we said above, this closed system model was designed to help urban administrators examine their cities and to project any given condition into the future to see possible implications of conditions over which they might have some control. It did not consider flows into or out of the city (such as water, air, pollutants, or people), because it was felt that the urban administrators had no control over such flows. The unit of analysis is the urban administrative system over which the administrator has direct control. It is true that administrative boundaries do not always conform to actual urban built-up areas. Since the 1980s for example, Chinese urban administrative boundaries extend far beyond the city and include large rural areas. Cities in Java, Indonesia are typically "underbounded," with the built up areas extending beyond the city boundaries. In Indonesia's outer islands cities are, as in China, typically "overbounded." Nonetheless, since we were concerned with the administrators' problems, it appeared best to work within the administrative boundaries of the city.

It is also, however, important to conceive of the urban system as an open system, with flows into and out of it. For this we turn to Fig. 2.2.

In the open systems model, we conceive of three types of inputs and outputs: Physical, Symbolic and Human. One of the major questions this type of model raises is the units of analysis. For water, we can easily identify the watershed as the boundary of the city, though they often cross national boundaries and present far more serious political problems. We can also think of an "air shed," though here, of course, we usually cross national boundaries. The air shed of Japanese and Korean cities includes the north of China. And the 1986 Chernoble disaster suggests such air sheds can be very large and highly complex. One could also think of a "food shed," indentifying the source of the majority of food that enters the city. Here is an area that has been expanding rapidly over the past half century. And, of course for such symbolic inputs the environment can include the entire world. It is relatively easy to conceive of the human inputs and outputs, since human migration is clearly a major force in any city. Migration is one of the major source of growth of the city's population and both the quantity and sources are fairly well documents.

This only illustrates how such a modeling exercise might proceed. We do not suggest that this is the appropriate model for examining and promoting sustainable cities. It only begins to identify the various environments, or active elements of a city. I believe it is especially important to identify the Social System of the city, or what we called the Social, Political, Economic and Cultural system. As I said above, I do not believe we yet have a suitable taxonomy or objective measure of these systems that speaks to the issue of sustainability. But consider one possibility. It is clear that cities, or Social Systems, differ along at least two dimensions that must be considered important for sustainability: the drive for economic development, and the concern for collective welfare. Consider the following four-fold table, which lays out the two dimensions as dichotomous attributes: high and low (Table 2.1). Considering both as variables would be better, but this simpler table will serve to introduce the analytical process we are considering. We can suggest certain countries at different times to indicate the scheme we have in mind.

In 1965 Japan gave highest priority to economic development, and serious pollutants like the mercury that produced Minamata Disease were ignored and covered up, sometimes violently. Today, Japan's emphasis on economic development may have waned a bit, but the issue of collective welfare is given much higher priority. In 1958 and 1967 China was pursuing a revolutionary dream and neither collective welfare nor economic development was considered important. These were times of the "Red," not the "Expert." China in 2005 shows especially strong emphasis on economic development and increasing awareness of the environmental costs of this emphasis. Thus the emphasis on collective welfare is gaining ground. India in the 1970s still exhibited the strong emphasis on state control and Socialism that placed welfare ahead of economic development. Since the economic reforms of the 1990s, economic development has gained higher standing. Myanmar may be said to have little interest in development or welfare since roughly 1962.

Again, this serves only to suggest how this taxonomy might be used to classify social systems on their impact on sustainability. It will be possible to consider each dimension as a variable and find objective indicators for each country at different points in time. Then it will be possible to develop objective indicators for all of the urban subsystems and assess their connections. From this forward projections under well defined assumptions can be made to assess future sustainability.

Table 2.1 Schematic diagram of The Social Political Economic Cultural Systems Relevant for Sustainable Development

	Collective welfare low	Collective welfare high
Ec Dev High	Japan 1965	China 2005
		India 2005
		Japan 2005
Ec Dev Low	China 1958/1967	India 1970
	Myanmar	

Different future scenarios can be run under different assumptions to understand the impact of various policy changes. It would be especially useful if RHIN could turn its attention to the further development of a taxonomy and set of variable measures to capture this important determinant of sustainable development. This would require more of the interdisciplinary research for whose promotion RHIN has already demonstrated considerable skill.

2.6 Conclusion

Here I have recorded my own appreciation of the excellent work of the RIHN Subsurface Environment Project. Discovering the condition of the unseen environment must be ranked as a major scientific advance. I have also noted that the cities studied are part of a massive movement of people, the urbanization of Asia. This has been far more rapid and far more massive than any past movement of peoples. This makes this large urban environment a highly critical one. As many have observed, the struggle for sustainability in the world will be won or lost in the cities. If that is true, a great deal of that struggle will take place in the Asian urban scene.

This makes it useful to consider some other urban environments suggested by the past studies of the subsurface environment project. I have listed some of these, from the larger social system of the city to the impact of permeable surfaces on the city's hydrology, to the increasing heat cities are generating, to the subsurface flows of pollutants. This has led me to make two suggestions of next steps.

First this project could develop a relatively standardized protocol for assessing the condition and health of a city's subsurface environment. This could be especially useful for international organizations like Habitat, UNEP, UNDP and the regional development banks.

Second, I suggest a somewhat larger next step: the dynamic modeling of urban systems to examine and help promote the development of sustainable cities.

References

Chandler T, Fox G (1974) 3000 Years of urban growth. Academic Press, New York

Flieger W (2000) Cebu City: heart of the Central Philippines. In: Ness and Low, pp 149–174

Huang S, Taniguchi M, Yamana M, Wang C.-h (USA, Japan, Taiwan, ROK) (2009) Detecting urbanization effects on subsurface and subsurface thermal environment – a case study of Osaka. In: Taniguchi M, Burnett W, Ness G (eds), pp 3142–3152

Jago-on KAB, Kaneko S, Fujikura R, Imai T, Matsumoto T, Zhang J, Tanikawa H, Tanaka K, Lee B, Taniguchi M (Japan, Republic of Korea) (2009) Urban and subsurface environment issues: an attempt at DPSIR model application in Asian Cities. In: Taniguchi, Burnett, Ness (eds), pp 3089–3104

Kataoka K, Matsumooto F, Ichinose T, Taniguchi M (Japan) (2009) Urban warming trends in several large Asian cities over the past 100 years. In: Taniguchi M, Burnett W, Ness G (eds), pp 3112–3129

Lee YW, et al (2009) Nutrient inputs from submarine groundwater discharge (SGD) in Masan Bay, an embayment surrounded by heavily industrialized cities, Korea. In: Taniguchi M, Burnett W, Ness G (eds), pp 3181–3188

Lieberman V (2003) Strange parallels: Southeast Asia in global context. Cambridge University Press, Cambridge

McCoy A (2009) An anarchy of families: state and family in the Philippines. University of Wisconsin Press, Madison

Ness GD, Low R (eds) (2000) Five cities: modeling Asian urban population environment dynamics. Oxford University Press, Singapore

Ness GD, Talwar P (2004) Asian urbanization in the new millennium. Marshall Cavendish, Singapore

Taniguchi M, Burnett W, Ness GD (2009a) Human impacts of the urban subsurface environments (special issue). Sci Total Environ 407:3073–3238

Taniguchi M, Shimada J, Fukuda Y, Onodera S-i, Kaneko S, Yoshikoshi A (Japan) (2009) Anthropogenic effects on the subsurface thermal and ground water environments in Osaka, Japan and Bangkok, Thailand. In: Taniguchi M, Burnett W, Ness G (eds), pp 3153–3164

UN (2005) World urbanization prospects, the 2005 revision. United Nations, New York

UNFPA (2007) State of the world population: unleashing the potential of urban growth. UNFPA, New York

Worldwatch Institute (2007) State of the world 2007: our urban future. W.W. Norton and Company, New York

Yamano M, Goto S, Miyakoshi A, Hamamoto H, Lubis RF, Monyrath V, Taniguchi M (2009) Reconstruction of the thermal environment evolution in urban areas from underground temperature distribution. In: Taniguchi M, Burnett W, Ness G (eds), pp 3120–3128

Yoshikoshi A, Adachi I, Taniguchi T, Kagawa Y, Kato M, Yamashita A, Todokoro T, Taniguchi M (2009) Hydro environmental changes and their influence on the subsurface environment in the context of urban development. In: Taniguchi M, Burnett W, Ness G (eds), pp 3105–3111

Chapter 3
Urban Development and Water Environment Changes in Asian Megacities

Akihisa Yoshikoshi

Abstract Since their growth into modern cities, Asian megacities have seen a change in their water environment due to such undertakings as the reclamation of their regional streams, rivers, lakes and ponds, and commencement of large-scale groundwater withdrawal projects. As a result, so-called water environment issues emerged in many megacities. A time-series analysis of the process of their emergence reveals that the earlier a city developed, the earlier water environment issues emerged. Accordingly, one can well expect that a city that is currently demonstrating remarkable growth might see in the near future an emergence of water environment issues similar to those in other cities that developed in earlier days. In this research, the urban development processes in the Asian megacities and resulting changes in their water environment were discussed, and the water environment issues that have emerged as a consequence were sorted out. As a result, it has been brought to light that the cities that developed early also saw water environment issues emerge early, but a good part of them are now in the process of being solved. This research stopped short of examining any specific actions taken to address those issues, but a remaining challenge is to apply in an effective fashion the approaches and experience of those cities to other megacities. As a city develops, changes occur to its water environment, such as a decrease in the area of its waters. Such changes would result in reduced groundwater recharge or might also diminish water retention and other functions served by the surface ground, possibly making it vulnerable to floods. It is a very important task to assess what changes in water environment lead to water environment issues to what extent, but there is still very little understanding of these questions. Presumably, a challenge remains for us to answer them in the future by studying a particular megacity.

A. Yoshikoshi (✉)
Department of Geography, Ritsumeikan University, 56-1 Tojiin Kitamachi,
Kita-ku, Kyoto 603-8577, Japan
e-mail: ayt03221@lt.ritsumei.ac.jp

M. Taniguchi (ed.), *Groundwater and Subsurface Environments: Human Impacts in Asian Coastal Cities*, DOI 10.1007/978-4-431-53904-9_3, © Springer 2011

3.1 Introduction

3.1.1 Objective

Since their growth into modern cities, Asian megacities have seen a change in their water environment due to such undertakings as the reclamation of their regional streams, rivers, lakes and ponds, and commencement of large-scale groundwater withdrawal projects. As a result, so-called water environment issues, i.e., lower groundwater levels, salinization of groundwater and land subsidence, emerged in many megacities. A time-series analysis of the process of their emergence reveals that the earlier a city developed, the earlier water environment issues emerged. Accordingly, one can well expect that a city that is currently demonstrating remarkable growth might see in the near future an emergence of water environment issues similar to those in other cities that developed in earlier days. If effective steps are taken now, before it is too late, the water environment issues to emerge in Asian megacities in the future may turn out to be different from those in the past. By the way, although there are the side of water resources and environmental impacts in the water environment issues of urban area, the author will deal with the environmental impacts about groundwater, mainly.

In this paper, cities that have followed different paths of development are examined: Tokyo, Osaka, Seoul, Taipei, Bangkok, Jakarta and Manila. Note, however, that those cities may not be regarded merely as administrative units like municipalities but are also viewed on a larger scale that covers their greater metropolitan regions. These megacities are the research subjects in "Human Impacts on Urban Subsurface Environments," a research project at the Research Institute for Humanity and Nature that the author is also involved with.

As mentioned above, urban development occurred earlier in cities like Tokyo and Osaka, and the resulting water environment changes and water environment issues emerged earlier there as well. These cities were able to take certain measures to address their water environment issues which, while insufficient, did indeed bear reasonable fruit. Presumably, a series of such occurrences has been experienced first in Tokyo and Osaka, followed by Seoul and Taipei, then Bangkok, and then Jakarta and Manila, in that order, as far as the cities discussed in this paper are concerned. The exact same types of water environment issues that emerged in Tokyo or Osaka would not likely emerge in other megacities: they each have a different natural environment and impacts on their respective water environments differ in both quality and quantity. Very similar water environment issues may, however, still emerge.

Viewed from the perspective described above, organizing and examining in chronological order the urban development processes, water environment changes, water environment issue emergence and other related matters in Asian megacities appears to be a task on which researchers in Japan should take the initiative. If done successfully, this could lead to creating proactive solutions to water environment issues expected to arise in Asian megacities, before they become serious.

There have been a good number of studies in the past on the subject of urban development in Asian megacities (e.g., Tasaka 1998; Nakanishi et al. 2001; Miyamoto and Konagaya 1999; Douglas and Huang 2002) and water environment issues have also been raised from various disciplines (e.g., Liongson et al. 2000; Porter 1996). While there is a study that compares two cities, namely, Tokyo and Bangkok (Matsushita 2005), virtually no research work has ever attempted a time-series examination of cities as many as those covered here. In this sense, writing this paper presumably carries some significance. This paper also serves the role of providing an overview of the megacities on which other papers of this book are to be based.

3.1.2 Methodology

To begin with, the author sought to define when the seven Asian megacities became modern cities and then gained an understanding of the process of their urban development since that time up until now. The task of deciding where the starting point of a certain city becoming a modern city should be set is rather challenging, given the variation between cities in terms of their own history as well as the socioeconomic conditions of their respective countries. Any time point so set may also not necessarily be directly relevant to the development process of the city as it is today. In the case of Tokyo and Osaka, for instance, such a base time point was set at 1868, the year of transition from feudal society to modernization society. Thus, in this paper, the author chose the time point at which the socioeconomic conditions of a given country (or city) changed dramatically and after which the city's spatial structure would start changing, and defined such time point as its beginning. Note that a somewhat different approach is used in the aforementioned project at the Research Institute for Humanity and Nature, which focuses on the period roughly after 1900 in order to align the time frame.

The next step involves an overview of urban development processes. Including individual discussions on each of the seven megacities, however, would only serve to obscure this paper's focus and is difficult due to constraints on the available space. On top of that, urban development processes of some of the cities have been explained in a paper by the author and others (Yoshikoshi et al. 2009). Therefore, this paper avoids giving a detailed account of urban development processes to the extent possible, and instead illustrates spatial development stages by a model-based approach.

Following that, water environment changes will be discussed and water environment issues overviewed for each city by presenting land use data in three stages of development process (*circa* 1930, *circa* 1970 and *circa* 2000). Based on such an understanding, observations will be offered as to any relations between urban development processes and the reality of water environment changes as well as water environment issues. The chapter will then be concluded with some suggestions from the author about an urban development process and water environment issues.

3.2 Beginning and Subsequent Development of the Megacities

Conceptually, the development processes of the seven megacities can be comprehended as shown in Fig. 3.1. To start with, the cities can be divided into two types when a focus is placed on what forms their core in terms of spatial aspects of urban development: cities walled with ramparts (Seoul, Taipei, Bangkok, Manila and Jakarta) and cities with a castle but not surrounded by ramparts (Tokyo and Osaka). In the former, the city developed in the area surrounded by clear boundary called rampart. In the latter, the city developed not in the area surrounded by clear boundary (rampart), but in the area centering on a castle. Both have the difference as shown above. The cities would later expand beyond ramparts, which would eventually be taken down in many cities (although they are preserved in Manila), with only some gates preserved up to the present time. Urban expansion would then further continue out to surrounding areas, which can also be divided into two types in terms of what kind of land was converted for development purposes: farmland and forestland. This is not as clear a distinction as the existence of ramparts, however.

Whether satellite cities have formed in surrounding areas is another factor that divides the megacities into two categories, but the timing of satellite city formation also varies from city to city. Bangkok and Jakarta have seen no obvious satellite city formation. These megacities have developed through the expansion of urban areas into their surrounding areas. Even in cities with satellite city formation, subsequent urban development and expansion eventually made it difficult to distinguish these cities from those without satellite cities in terms of landscapes, resulting in the formation of so-called conurbation cities. Figure 3.1 illustrates all these aspects in chronological order, with the most recent period at the bottom.

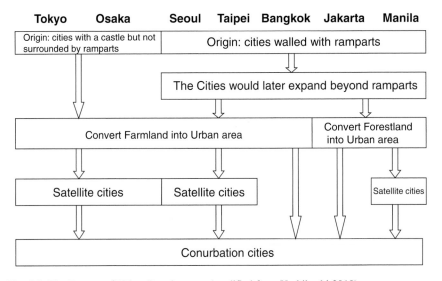

Fig. 3.1 The Process of Urban Development (modified from Yoshikoshi 2010)

Table 3.1 The data on metropolitan region on Asian megacities

Megacities	Population (10,000 persons)	Metropolitan area (km²)	Population density (person/km²)
Tokyo	3,425	7,835	4,350
Osaka	1,725	2,720	6,350
Seoul	1,950	1,943	10,050
Taipei	650	440	14,750
Bangkok	800	1,502	5,350
Jakarta	2,060	2,720	7,600
Manila	1.915	1,425	13,450

Source: Demographia (http://www.demographia.com/)

While a description of urban development processes in this paper should revolve around Fig. 3.1, doing so would be a little too complicated. Alternatively, therefore, the cities are divided into the four groups below in an attempt to provide an overview and present spatial development models for each group on the basis of urban development stages.

As part of research in our project at the Research Institute for Humanity and Nature, the land use data on the subject megacities over three periods was converted into GIS data. More specifically, GIS was used to create square-grid land use maps on the basis of 1:50,000-scale topographical maps from the three respective periods. In this process, roughly 500-meter square grids were drawn on topographical maps, with each grid showing the most prominent way in which the land was used in the area. Therefore, this method would preclude any comparative analysis using detailed figures, since minor land use that cannot be expressed by this approach would be ignored and because of variations in the topographical maps, on which the analysis would be based, depending on the country or the year in which they were created.

The data on the greater metropolitan regions of Asian megacities (greater metropolitan population, greater metropolitan area, population density) is as shown in Table 3.1, for the purpose of comparison the sizes of recent greater metropolitan regions.

3.2.1 Tokyo and Osaka

As was already mentioned, the starting point of Tokyo and Osaka as modern cities is set at the Meiji Revolution of 1868. The Meiji Revolution was a period of major innovations and modernization in Japan, when the shogunate regime collapsed and Japan-style capitalism and nation-building based on modern imperialism began.

Tokyo and Osaka are both situated on coastal plains and adjacent to rivers. Tokyo, in particular, had vast farmland to its back consisting of rice paddies and crop fields, allowing physical room for the city to expand substantially. The two cities used to be castle towns in pre-modern times but were not surrounded by

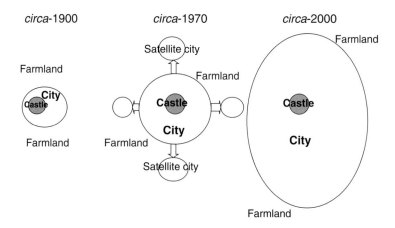

Fig. 3.2 Urban development model of Tokyo (reprinted from Yoshikoshi 2010)

ramparts. They also shared a common feature of having inside them a large number of moats, streams and other types of waters that were used, among other ways, to carry people and transport goods.

Later on, Tokyo's development began mainly from areas around the castle, as is illustrated in Fig. 3.2. Of the two cities, Tokyo, in particular, suffered devastation: it was utterly destroyed by the Great Kanto Earthquake in 1923 and then by air raids in the last phase of World War II, but thereafter made an astonishing recovery. Tokyo and Osaka took in a large population coming from all around Japan not only in their urban areas but also by forming satellite cities in their suburbs. The inner city areas and their satellite cities were connected by railways that radiated out from the center. The cities expanded rapidly during the period of high economic growth starting in the mid-1950s, resulting in a large number of satellite cities growing in their suburbs as residential areas for those commuting to the city centers by train, etc. Further growth that followed led to conurbation of existing medium- and small-sized cities in the vicinity and satellite cities, which entailed across-the-board urbanization that covered all those areas. The aforementioned Fig. 3.2 is a model representing these actual developments. Osaka is represented by the same model as Tokyo since both cities essentially followed the same path of development, with the greatest difference being that Osaka suffered no damage from the Great Kanto Earthquake. As can be seen from Table 3.1, the size of Tokyo as a city is truly world-class.

Figure 3.3 illustrates changes in land use in Tokyo. Residential lands existed only in the proximity of coastal areas around 1930. However, by roughly 2000, they were seen to extend from the areas along Tokyo Bay to the inland areas. In contrast, it is clear that farmland followed a shrinking pattern. Likewise, changes in land use in Osaka are shown in Fig. 3.4. This figure shows, just as tellingly as in the case of Tokyo, how farmland was converted to provide more and more residential land.

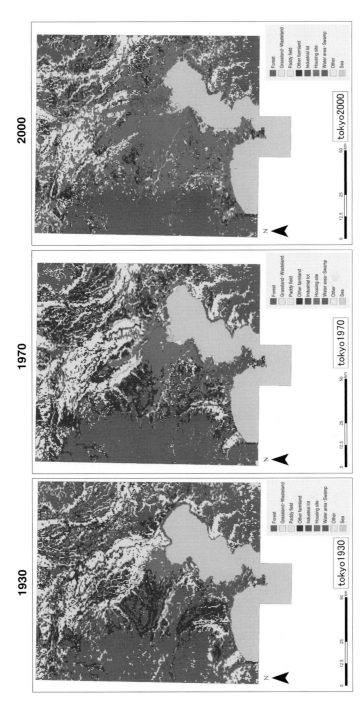

Fig. 3.3 Land use of Tokyo (reprinted from Yoshikoshi 2010)

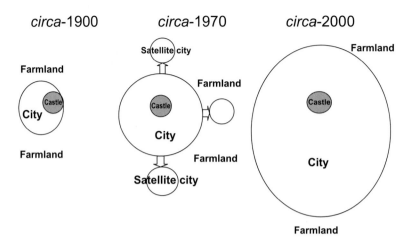

Fig. 3.4 Urban development model of Osaka (reprinted from Yoshikoshi 2010)

3.2.2 Seoul and Taipei

After the Japan-Korea Annexation Treaty was signed in 1910, Seoul was renamed as Gyeongseong-bu and the Japanese Governor-General of Korea was established there. The author defined this point in time as the beginning of Seoul as a modern city. Likewise, Taipei as a modern city was deemed to have begun when it came under Japanese rule with the establishment of the Japanese Governor-General of Taiwan in 1895.

Seoul and Taipei are both cities originating in a relatively narrow basin area. At the onset of modern-city formation, Seoul had crop fields and also some forestland in its surroundings, while Taipei had an extensive sprawl of farmland, such as rice paddies and crop fields, in its surroundings. Although both cities were enclosed by ramparts, the area within them was particularly small in Taipei, with their construction dating back to 1875, which is also substantially new in comparison to other cities. The ramparts were eventually removed as both cities developed beyond their borders, with only some gates remaining at present. Seoul expanded from the basin on which the city center lies, out to the proximity of the Hangang River in the south, which has now grown to an area serving the function as the new city center. Satellite cities, including Incheon and Suwon, developed in the west and the south of Seoul. Meanwhile, Taipei achieved rapid growth in relatively recent times by having satellite cities form in areas in the south to the west across the Danshui River, major examples of which are the cities of Yonghe, Banqiao and Sanchong (Selya 1995). Thus, the development process of Taipei is very similar to that of Seoul. For that reason, the charted model for Seoul (Fig. 3.5) is used as one representative of Taipei as well.

As can be seen from Table 3.1, what is characteristic of Taipei is that its population density is the highest among the subject cities despite its small population because its urban area is very small.

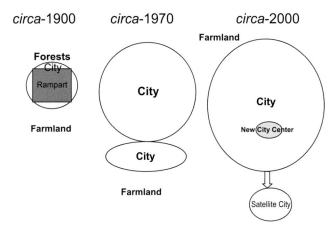

Fig. 3.5 Urban development model of Seoul (reprinted from Yoshikoshi 2010)

Changes in land use in Seoul are shown in Fig. 3.6. Around 1920, rapid expansion of the city's core, which used to be located to the north of the Hangang River, can be observed. However, the area had failed to expand throughout due to limitations posed by mountains and other terrain in its surroundings.

3.2.3   Bangkok

In Bangkok, the *coup d'etat* that broke out in 1932 (a bloodless revolution) resulted in a transition from absolute monarchy to constitutional monarchy. The author defines this point in time as the beginning of Bangkok as a modern city. The city of Bangkok originated from an area surrounded by ramparts and canals, with Rattanakosin Island in the Chao Phraya River at its center. Subsequently, the urban area expanded further as more canals, etc. were built. Such expansion initially followed the Chao Phraya River and then successively expanded out to the east and the south with the construction of roads.

Areas surrounding Bangkok were a vast delta with an expanse of rice pad sea level.

Since 1980, Bangkok ballooned by attracting a sizable population from rural regions in its surroundings. Bangkok's development proceeded not through satellite city formation, but through successive expansion of its urban area by means of having its surrounding farmland converted. The Bangkok Metropolitan Area (BMA), which covers Bangkok and its surrounding areas, thus further grew to entail the formation of the Bangkok Metropolitan Region (BMR).

A model designed to represent the urban development in Bangkok is as shown in Fig. 3.7. It shares a common feature with Jakarta in that it is void of satellite city development, but can still be regarded as having a unique pattern in that there has been no obvious formation of a new city center as was the case of Jakarta.

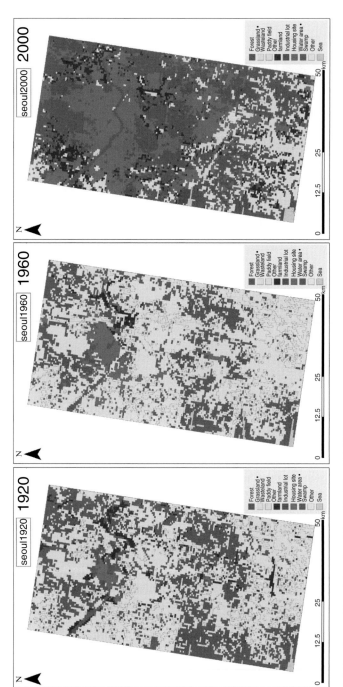

Fig. 3.6 Land use of Seoul (reprinted from Yoshikoshi 2010)

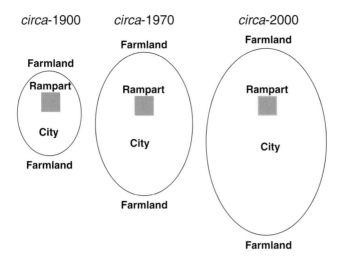

Fig. 3.7   Urban development model of Bangkok (reprinted from Yoshikoshi 2010)

3.2.4   *Jakarta and Manila*

Jakarta's beginning as a modern city was set at the period of the "Liberal Policy" implementation during the 1860s. While this period may seem early, relative to the other megacities considered, it was selected in consideration that it was at that time when various forms of infrastructure worthy of a modern city were built in succession. Jakarta has a history of foreign rule by Portugal, the Netherlands and Japan. Meanwhile, the beginning of Manila as a modern city was defined at the period of its cession from Spain to the USA in 1898. Both cities share a common feature of having originated in a ramparted area and are also quite similar in that their development involved new city center formation and transitions. In Manila's case, Metro Manila was formed through the formation of many cities in its environs and the scale of those cities now far exceeds that of Manila. Jakarta, in contrast, grew with no obvious satellite city formation but rather by having its urban area expand into its surroundings and absorbing a large population into its inner city. Today, the Special Capital Territory of Jakarta (DKI Jakarta) exists as a result of that process. The greater metropolitan Jakarta that includes Jakarta's surrounding areas is called Jabotabek. Although their formation mechanisms are somewhat different, both megacities are also in a similar situation in that they have a large number of slums within their urban areas.

Two additional similarities between these cities are worthy of note. Both are located on or in close proximity to coastal plains. Jakarta had its urban area sprawl to a southern region where elevations above sea level increase gradually, whereas Manila's equivalent expansion took place on the isthmus-like lowland sandwiched between Manila Bay and Lake Laguna. Another common aspect is the former existence of forests or plantations in the regions into which they expanded their urban areas.

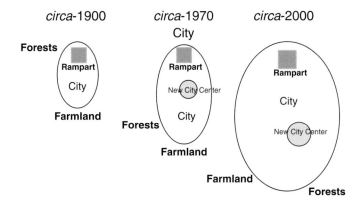

Fig. 3.8 Urban development model of Jakarta (reprinted from Yoshikoshi 2010)

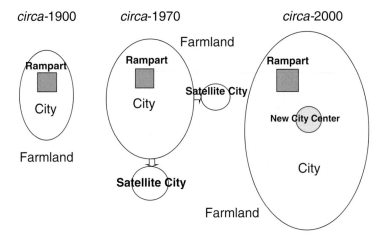

Fig. 3.9 Urban development model of Manila (reprinted from Yoshikoshi 2010)

The subsequent development proceeded as shown in Fig. 3.8 in the case of Jakarta and in Fig. 3.9 in the case of Manila. Although their respective development processes are quite similar, a different model was designed for each because they are slightly different in that, at their stage of development in 1970, Manila saw the formation of its satellite cities, whereas Jakarta had new city centers formed and transitioned.

Table 3.1 also shows that a characteristic of Manila is its having a high population density. Changes in land use in Manila are illustrated in Fig. 3.10. It clearly shows the process of Manila's urban area expanding to the east at a gradual pace and then rapidly to the south and the north.

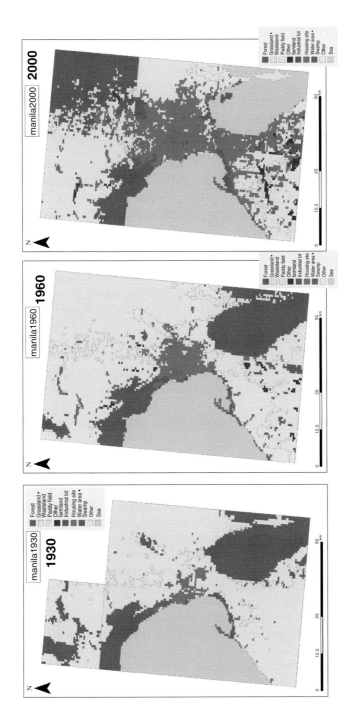

Fig. 3.10 Land use of Manila (reprinted from Yoshikoshi 2010)

3.3 Water Environment Changes and Water Environment Issues

Grouping Asian megacities into several categories on the basis of water environment changes and water environment issues is no easy task, given that such changes and issues are not only simply related to urban development, but are inherently connected with the natural environment of the region in which the city is located. Accordingly, a description of these aspects will be given in relation to each megacity, while it may admittedly appear a little digressive.

In this context, water environment changes refer to areas of water, including channels, streams, rivers, lakes, ponds, and marshlands, changing to roads or other urban-type land uses. Although it is naturally expected for water areas and marshlands to diminish with urban development, such changes in land use may not always be reflected accurately when land uses are calculated by a grid-square approach based on topographical maps as described earlier. With reference to GIS data, outcomes are relatively good in the cases of Tokyo, Seoul, Jakarta, Manila, etc., but appear to leave some issues for other cities. One way to complement this would be to use large-scale maps produced in different years to compare waters and other areas on them. This approach will be used later to describe the example of Bangkok.

Also note that in this chapter, water environment issues are almost entirely limited to groundwater level fluctuations, land subsidence, groundwater salinization, and water pollution; furthermore there are gaps present in the data, as all figures could not be collected for all the megacities.

3.3.1 Tokyo

1. Water environment changes
   The area of its waters and marshlands dropped from 336 km^2 around 1930 to 173 km^2 (GIS data) by around 2000. Waters and marshlands diminished because of, among other things, their conversion to roads, etc. and culverting of streams. Since increased river flows as a result of urbanization were seen to lead to flood damage in recent years, a series of new measures have started being taken, such as stream or river improvements together with underground drain construction.
2. Groundwater level fluctuations
   Since a restriction on groundwater pumping was put in place in 1961, groundwater levels have been on the rise. Prior to 1970, the amount of water pumped up in Tokyo per day used to be 1.5 million m^3, but has recently declined down to 400,000 m^3. This, in turn, actually caused groundwater levels to rise too high and has thus generated new issues, such as water seepage into underground structures as well as deformation and surfacing of them due to water pressure (Aichi and Tokunaga 2007).

3. Land subsidence

   Land subsidence had already been occurring since the 1880s when monitoring was begun, but the issue was brought to light only after the 1920s. The trend ceased temporarily during World War II but again became noticeable in post-war days and continued on until around 1970, when the water pumping restriction started to be proven effective (Endo et al. 2001). Viewed by area, land subsidence took place mainly in areas from the Tokyo Bay coastal area to the alluvial lowland. Thanks to the groundwater pumping restriction that was subsequently applied to most areas, the land ceased to subside after 1970. On the diluvial plateau where groundwater pumping still continued, however, localized subsidence was observed (Endo and Ishii 1984). This led to the appearance of so-called "zero-meter zones," meaning areas of which elevation is at or below sea level, on the alluvial lowland, including the proximity of the Arakawa River's mouth where a large cumulative subsidence was recorded. It therefore became necessary to construct coastal levees, etc. to prevent tidal wave damage.

   With the cessation of land subsidence, the ground was uplifted, albeit very little, principally in areas with a large cumulative subsidence; one area had a cumulative uplift of as much as 16 cm (Endo et al. 2001).

4. Water salinization

   When groundwater levels dropped sharply, coastal areas experienced groundwater salinization. This issue, however, has been put to rest in recent years because groundwater levels rose and groundwater is no longer used. Of an additional note, the problem of accidents caused by oxygen-deficient air, which had occurred frequently on underground and other construction sites between the 1950s and the 1970s, was likewise essentially solved as a result of the groundwater level rise (Endo and Ishii 1984).

5. Water pollution

   Non-point source pollution from nitrate-nitrogen and other compounds used to be found in Tokyo's groundwater, though the recent trend has shifted to point source pollution (from heavy metals and the like) from factories and other sites.

3.3.2   Osaka

1. Water environment changes

   Osaka used to have a large number of small channels and canals, but many of them were reclaimed and were converted to roads.

2. Groundwater level fluctuations

   The first observation of groundwater levels in Osaka was conducted in 1939. The levels rose temporarily during World War II but fell sharply again as a result of pumping of a large quantity of groundwater after the war (Nakamachi 1977). Since a restriction on groundwater pumping was enacted in the late 1950s, groundwater levels followed an upward trend, which eventually regained wartime levels in 1965 and continued rising thereafter as well. As with Tokyo, this has generated problems of water seepage into underground structures as well as surfacing of them in Osaka.

3. Land subsidence

   Thanks to the restriction on groundwater pumping imposed by new laws and local bylaws, land subsidence ceased in Osaka around 1970. Subsequently, in some places, the ground has swelled by several centimeters in a cumulative measurement. Land subsidence measurements in Osaka were large in the Higashi-Osaka (eastern) region as well as the coastal area of Osaka Bay and the lowland in the vicinity of the Yodogawa River (Nishigaki 1988). Meanwhile, even in the heart of Osaka, only minor land subsidence was observed in areas such as the Uemachi Plateau, on which Osaka Castle and other structures are built. This suggests that a characteristic of land subsidence in Osaka is that it patently reflects geographical and geological conditions.

4. Water salinization

   Water salinization emerged as an issue in coastal regions, etc. in the period of lowering groundwater levels but, thanks to increased groundwater levels and a sharp drop in groundwater use, it is no longer a water environment issue.

5. Water pollution

   Osaka has been following a trend very similar to that of Tokyo in that point source pollution from organic compounds is now dominant.

3.3.3 Seoul

1. Water environment changes

   The trend of small- and medium-sized streams being converted into roads, etc. was observed in Seoul as well. This phenomenon can also be seen from the fact that the area of its waters dropped from 318 to 247 km^2 in GIS data.

   Given that a general trend in Asian megacities is that waters, such as streams, rivers, lakes and ponds, are turned into roads, etc., the restoration of the Cheonggye River in Seoul was a feat worthy of special mention. The Cheonggye River is a small urban stream of 11 km in length that flows eastwards within the Seoul Basin and then joins the Hangang River. In order to address Seoul's traffic volume, which was rising with the progress of urbanization, a construction project for culverting the Cheonggye River was per formed in the early 1950s, which was followed by the construction of an elevated road above the culvert. The subsequent deterioration of the environment triggered growing calls for waterfront restoration and, by 2005, the restoration project was completed (Haruyama 2006) to a successful return of waterfront sceneries in the heart of downtown Seoul, as can be seen in Fig. 3.11. This is a rare example among all Asian megacities.

2. Groundwater level fluctuations

   An increase in the quantity of groundwater getting pumped up entailed lower groundwater levels. Groundwater levels in Seoul dropped by 60 cm in 6 years in

Fig. 3.11   Cheonggye river in the city of Seoul

the recent past. In addition to the increased quantity of groundwater pumping, the fact that the city's ground surface is now covered with impervious materials is suspected to be another reason for this.

3. Land subsidence

As Seoul sits on a granite basement, no land subsidence would occur where sedimentary layers are thin. In areas with relatively thick sedimentary layers, however, land subsidence has indeed taken place and caused cracking of building structures and other damage. Land subsidence problems similar to those experienced in other megacities have emerged in Seoul's coastal regions as well, including the satellite city Incheon.

4. Water pollution

Recently in Seoul, the state of bacterial pollution (especially E. coli) and water pollution (from organic solvents and heavy metals, etc.) is reportedly quite serious. While it is a case unique to Seoul, some also point out that subway air is contaminated with radon that is naturally emitted from its basement granite.

3.3.4   Taipei

1. Water environment changes

In Taipei, the area of waters was decreased by the straightening of river channels rather than the reclamation of streams, rivers, etc. This, however, does not appear to be reflected accurately in figures in GIS data.

2. Groundwater levels

Groundwater levels once dropped as a result of a large quantity of water being pumped up, but then started rising around 1972 following a restriction on

groundwater pumping enacted in 1968. In Taipai and its satellite city Sanchong, groundwater levels rose by as high as 30 m between 1976 and 1994.
3. Land subsidence
   Sedimentary layers in Taipei's inner-city area are approximately 700 m thick to the basement. Due to this natural environment and the pumping of groundwater in large quantities, cumulative subsidence measurements since 1951 marked around 2.2 m in some areas with a higher subsidence, such as Taipei and the vicinity of Sanchong lying across the Danshui River from Taipei. After the 1968 restriction of groundwater pumping was set in place, subsidence measurements started dropping and are said to have now almost ceased. There has been a press report, however, that an annual subsidence has reached 2.4 cm and a cumulative subsidence 7.2 cm in 3 years in the recent past (2004–2006), which might suggest that subsidence has not ceased completely.
4. Water salinization
   Electrical conductivity of Taipei's groundwater measures roughly between 250 and 500 µS/cm, but it exceeds 750 µS/cm in some area in the east. Due to its distance from the coast, however, this has not entailed any serious consequences (Chen 2005).

3.3.5 Bangkok

1. Water environment changes
   Rivers and canals have been reclaimed in Bangkok. However, a time-series comparison of Bangkok's inner-city area between 1917 and 2004 using large- scale maps shows that the area of its waters has decreased substantially in recent years, as is illustrated in Fig. 3.12. In particular, it was found that there are many reclaimed pieces of land that used to be short canals directly connected to the Chao Phraya River, which flows southwards down the middle of the inner city, or that used to be canals on its right bank or in other areas.
2. Groundwater levels
   It was after the early 1950s that groundwater started to be pumped up in a large quantity in Bangkok. A daily pumping volume was approximately 8,000 m³ back then, but it jumped to as much as approximately 1.4 million m³ by 1982. As many houses and factories were built even out in the suburbs during the so-called economic bubble in Thailand after 1993, the groundwater pumping volume increased to eventually exceed two million m³ by 1999 (Noppadol et al. 2008).
   Up until around 1998, although public groundwater pumping dropped due to the establishment of the Groundwater Act in 1977, private groundwater pumping showed no sign of slowing down and actually even surged after 1990. Later, following the start of public water supply using surface water of the Chao Phraya River in 1981, the groundwater pumping volume has been on the decline. As an outright ban on groundwater pumping was set in place in 2004, the groundwater level decrease is now in the process of stabilizing.

Fig. 3.12   Water area of Bangkok (reprinted from Yoshikoshi 2010)

3. Land subsidence

In Bangkok, sedimentary layers are approximately 500 m thick to the basement. Land subsidence is now a serious water environment issue in Bangkok, especially in the vicinity of the inner city where particularly serious subsidence occurred. In some places, including an area east of the Chao Phraya River, the largest subsidence on record was 85 cm for the period from 1933 to 1978, 75 cm from 1978 to 1987 and 38 cm from 1992 to 2000. Where the largest subsidence was observed, a cumulative measurement of over two meters was recorded for the 1933–2002 period. While subsidence measurements are currently dropping in the inner city thanks to the groundwater pumping restrictions, marked subsidence is still progressing in the suburbs. This has caused a string of different types of damage, including cracks in building structures due to uneven sinking and the collapse of a pagoda in Wat Saket. Further still, the fact that Bangkok is situated on flat lowland is a cause of frequent floods, as poor drainage conditions occur in rivers and canals in rainy seasons. In particular, flood damage suffered from 1983 to 1985 was quite substantial. Flood prevention steps have therefore been taken, including levee construction and discharge pump station construction (Noppadol et al. 2008) as well as comprehensive flood control measures using a flood control basin (Matsushita).

4. Water pollution

Groundwater pollution is caused not only by drainage from factories (in the sugar industry, the pulp and paper industry, the rubber industry, etc.) but also by human and agricultural sewage. Pollution from chlorinated organic solvents has also become marked in recent years.

3.3.6   Jakarta

1. Water environment changes
   While Jakarta has no major river, it has extremely meandering, medium- and small-sized streams flowing within it. It can also be seen from GIS data that the area of its waters was slightly reduced by having their meandering parts straightened.
2. Groundwater levels
   In Jakarta, sedimentary layers are approximately 300 m thick to the basement. Since 1980, there has been a roughly 2.5-fold increase in the number of deep wells in DKI Jakarta, which coincided with an increase in the volume of underground water pumping. The pumping volume has now reached 750 million m^3 per year (Douglas and Huang 2002). Behind these figures is the reality that drinking water is supplied to only 20 percent of the people of Jakarta (Wada et al. 2006). As a consequence, groundwater levels have dropped considerably as well.
3. Land subsidence
   Land subsidence is now a serious water environment issue in Jakarta. There were areas in the northern part of its downtown that were found to have major land subsidence: between 1982 and 1991, it occurred on the west and east sides; between 1991 and 1997, particularly on the west side; and between 1997 and 1999, on the east side. Also after 1995, subsidence has been continuing at the yearly rate of 19 cm in Jakarta's western part and 11–13 cm in its central part. This has caused frequent occurrences of tidal wave damage in coastal regions. In particular, places like Muara Baru are flooded virtually on a daily basis (Hirose et al. 2001). Figure 3.13 is a scene from a flood in the vicinity of the Sunda Kelapa port that the author experienced firsthand in 2007. Possibly because flooding has become an everyday event for them, residents did not appear to be panicking.

Fig. 3.13 High tide disaster in the Vicinity of Sunda Kelapa

4. Water pollution
   Groundwater pollution is also serious in Jakarta, the most dominant form of
   which is bacterial and organic in nature. Heavy metal and oil pollution has also
   taken place near the Tanjung Priok port in the city's northeastern part.

3.3.7   Manila

1. Water environment changes
   That the area of Manila's waters has also decreased substantially can be judged
   from GIS data as well. For flood prevention purposes, improvement projects
   have started to be carried out in recent years to make the water flow better, as
   will be described later.
2. Groundwater levels
   Groundwater levels have declined to a significant extent far and wide in the city's
   urban area.
3. Land subsidence
   Land subsidence is currently taking place in the greater metropolitan Manila region,
   as a consequence of which an extensive part of it now sits below sea level.
      In Manila, floods also occur frequently not only due to urbanization but also
   because of the poor drainage characteristic of the flat lowland. Recently, floods
   have occurred near Lake Laguna as well. This prompted improvement projects
   to be carried out in the Napindan Channel, a floodway from Lake Laguna, and
   the Pasig River, as well as drainage pump station construction, etc. (Douglas and
   Huang 2002), to which Japan provides assistance. A coastal area called the
   Navotas District is a slum sitting on a landfill and serving as home to approximately 5,000 people, where there is a forest of shacks built on the water to avoid
   flood damage. Water environment issues have emerged in close connection with
   urban problems like this.
4. Water pollution
   Water pollution is also considerable not only in streams and rivers, or lakes and
   ponds, but also in the ocean and groundwater alike. Serious water pollution is taking
   place because sewage treatment facilities are inadequate, which has led to insufficient treatment of wastewater from general households, factories, farmland, etc.

3.4   Connection Between Urban Development
and Water Environment Changes

The observations in the preceding sections have revealed that there are time lags
between megacities in terms of timings of water environment issue emergence and
solution. On that account, the current state of water environment issues in the

Table 3.2 Current state of water environmental issues (modified from Yoshikoshi 2010)

| Cities | Water environmental issues (items) | | | | | Sewage service coverage rate (%) |
	Ground water level	Land subsidence	Ground water salinization	Ground water pollution	Total	
Tokyo	○	◎	◎	Δ	○	99.9
Osaka	○	◎	◎	Δ	○	99.9
Seoul	Δ	Δ	–	Δ	Δ	98.1
Taipei	○	Δ	○	–	Δ	60.1
Bangkok	Δ	×	–	×	×	50.0
Jakarta	Δ	×	–	×	×	1.0
Manila	Δ	×	–	×	×	5.0

◎ very positive, ○ positive, Δ somewhat problematic, × very problematic, – un-inquiring

respective cities was examined and sorted out for this paper. The results are as shown in Table 3.2. As the data used was not necessarily prepared on the same basis, elaborate comparison becomes difficult; therefore, the approach chosen by the author was to compare the degree of water environment issues on a scale of four 5: ◎ indicates very positive, ○ positive, Δ somewhat problematic, × very problematic and – un-inquiring. Take groundwater levels as an example. Tokyo and Osaka should really be given a score of ○, seeing that their groundwater levels are on the rise, but the score given is ◎ because, as mentioned above, they are faced with new issues arising as a result of such rise. One can clearly see from the table that water environment issues have been solved to a considerable degree in Tokyo and Osaka; further, it is demonstrated that solutions are under way in Seoul and Taipei, while signs of solution can be seen to appear in Bangkok. Discernibly, water environment issues in Jakarta and Manila are currently in the most serious condition. Note that the sewage service coverage rate figures presented in the table concern the cities in administrative terms, which are more limited than greater metropolitan regions. As mentioned above, the issue of groundwater pollution has yet to be solved in Tokyo and Osaka as well. As sewage systems are effective in the face of organic pollution but have no effect on heavy metal pollution, etc., new steps need to be taken in order to attain a positive state, even in the case of a city with almost 100-percent sewage service coverage. This observation provides a fairly clear picture of the state of water environment issues that the Asian megacities are currently faced with.

As a next step, an attempt will now be made to gain an organized time-series view as to how water environment issues have developed. Note that the water environment issues focused on here are limited to lower groundwater levels and land subsidence; these issues are examined to ascertain the timing at which any major change occurred to them. The development processes of the megacities are as illustrated in Sect. 3.2. In every city, it is only after entering into a period of high economic growth that certain steps are taken to address water environment issues. As solving any water environment issue requires a proportionate amount of money, an economic backing is imperative. Figure 3.14 illustrates how those factors are

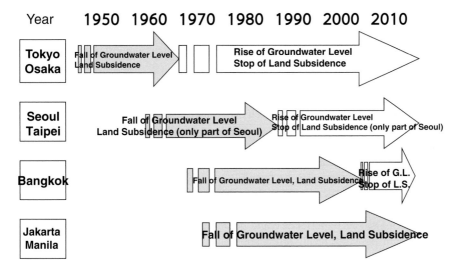

Fig. 3.14 Change of water environment issues (modified from Yoshikoshi 2010)

connected with each other. Viewed by city group, Tokyo and Osaka entered into a high economic growth period in 1955, during which they managed to get out of the most serious stage of their water environment issues. The economy of Seoul and Taipei grew rapidly between 1980 and 1985. It is from then on that their water environment issues have been in the process of being solved. Equivalent economic growth started in 1980 in Bangkok, but it is only very recently that signs of the city getting over its water environment issues came into view. In the cases of Jakarta and Manila, where a high economic growth period began in 1990, water environment issues are still in their serious stage. Thus, it has now become clear that every city group has gone through, or is going through, a similar experience with time lags of roughly 20 years.

3.5   Conclusion

In this paper, the urban development processes in the Asian megacities and result-ing changes in their water environment were discussed, and the water environment issues that have emerged as a consequence were sorted out. As a result, it has been brought to light that the cities that developed early also saw water environment issues emerge early, but a good part of them are now in the process of being solved. This chapter stopped short of examining any specific actions taken to address those issues, but a remaining challenge is to apply in an effective fashion the approaches and experience of those cities to other megacities.

In conclusion, an attempt will be made to examine which aspects of Tokyo's or Osaka's experience must be communicated to other cities. As an example, the case of Bangkok will be discussed. Bangkok has gone through its urban development and the emergence of water environment issues with a time lag of approximately 30 years compared to Tokyo or Osaka. Looking back at how it used to be in Tokyo and Osaka three decades ago, Japan was frantically trying to control its surging demand for water. What Tokyo and Osaka did back then was having water saved to the furthest extent possible. Specific examples were: charging a lot for water, promoting the reuse of water in factories, etc., and facilitating the broader use of water-saving appliances in factories and at home, etc. Efforts were also made to prevent leaks in waterworks and reduce unnecessary water use. Depending on the weather conditions, however, droughts took place and water-saving awareness needed to be raised by means of publicity. Now, 30 years on, the issues that were pending then have been solved to a considerable extent and Tokyo and Osaka are now water-saving-minded cities. It is unclear in what form Tokyo's or Osaka's experience could be put into action in, for instance, Bangkok, but it might be worth trying a few approaches in this light.

As a city develops, changes occur to its water environment, such as a decrease in the area of its waters. Such changes would result in reduced groundwater recharge or might also diminish water retention and other functions served by the surface ground, possibly making it vulnerable to floods. It is a very important task to assess what changes in water environment lead to water environment issues to what extent, but there is still very little understanding of these questions. Presumably, a challenge remains for us to answer them in the future by studying a particular megacity.

In megacities, however, a whole new issue has emerged in relation to water: floods. Global-scale changes in climatic conditions (e.g., global warming, sea level rise, so-called guerilla downpours) have resulted in local-scale flood damage. Megacities, however, remain virtually inept at dealing with this phenomenon. In all likelihood, this issue will also need to be brought into focus in the future.

Acknowledgement This study was funded by a research grant from a project at the Research Institute for Humanity and Nature entitled "Human Impacts on Urban Subsurface Environments" (Project Leader: Professor Makoto Taniguchi), and some data, including GIS data, was also provided by the same project. Moreover, cooperation of the member of "Urban Geography Group" was obtained in this study. I express my gratitude to all the support that was so provided.

References

Aichi M, Tokunaga T (2007) Groundwater environment issue transitions in urban areas and future prospects: an example of Tokyo lowland. Soil Mech Found Eng 55(8):5–8
Arai T (1996) Hydrological environment changes in Tokyo. J Geogr 105:459–474
Asada M (2008) Groundwater environment in Asian regions, 5. groundwater environment and waste treatment/soil pollution in Thailand. Soil Mech Found Eng 56(1):58–65
Berner E (1997) Defending a place in the city. Ateneo de Manila University Press, Quezon City, pp 1–34

Blusse L (1983) Dutch East India Company and Batavia (1619 to 1799): cause of the town's collapse. Southeast Asian Stud 21(1):62–81

Chen W-H (2005) Groundwater in Taiwan. Yuanzu Wenhua, Taiwan, 213

Douglas I, Huang, S-L (2002) Urbanization, East Asia and Habitat II. UN NGO Policy Series No. 2. IRFD-East Asia Network, Taiwan, p 326

Endo T, Ishii M (1984) Phenomena resulting from hydrogeology and groundwater level rise in the plain region of Tokyo prefecture. Appl Geol 25(3):11–20

Endo T, Kawashima S, Kawai M (2001) History of development and stabilization of "zero-meter zones" on lowland in the old downtown Tokyo. Appl Geol 42(2):74–78

Grijms K, Nas PJM (eds) (2000) Jakarta-socio-cultural essays. Kitlv Press, Leiden, p 349

Haruyama S (2006) An attempt to restore clear waters in an urban stream: a case of the Cheonggye River in the city of Seoul. Water Sci 289:29–44

Heuken ASJ (2000) Sumber-sumber asli sejarah. Jilid II, Jakarta, p 160

Heuken ASJ (2001) Sumber-sumber asli sejarah. Jilid III, Jakarta, p 128

Hirose K, et al (2001) Land subsidence detection using JERS-1 SAR Interferometry. In: Proc ACRS 2001, The 22nd Asian Conference on Remote Sensing, 5–9 Nov 2001, Singapore, pp 1–6

Iimi A (2004) Urbanization and infrastructure development in East Asia. JBICI Rev 20:4–25

Ishii M (1977) Kanto plain (Vol. 1) land subsidence in Tokyo. Soil Mech Found Eng 25(6):29–36

Lin J-C (2004) Natural disasters in Taiwan. Yuanzu Wenhua, Taiwan, 189

Liongson LQ, Tabios GQ III, Castro PPM (eds) (2000) Pressures of urbanization: flood control and drainage in Metro Manila. The UP-CIDS, Quezon City, p 105

Matsushita J (2005) Water environment transitions resulting from urbanization and countermeasures: a comparison between Tokyo and Bangkok, Laboratory of Regional Design with Ecology, Graduate School of Hosei University (ed) Cities and housing alongside the Chao Phraya River. pp 110–141

Miyakoshi A, Hayashi T, Marui A, Sakura Y, Kawashima S, Kawai M (2006) Assessing groundwater environment changes on Tokyo lowland based on underground temperature. Appl Geol 47(5):269–279

Miyamoto K, Konagaya K (1999) Big cities in Asia[2]. Nippon Hyoronsha, Jakarta, p 370

Nakamachi H (1977) Land subsidence in the Osaka plain. Soil Mech Found Eng 25(6):61–67

Nakanishi T, Kodama T, Niitsu K (eds) (2001) Big cities in Asia[4]. Nippon Hyoronsha, Manila, p 269

Nishigaki Y (1988) Land subsidence. Soil Mech Found Eng 36(11):27–32

Noppadol P, Otsu H, Nutthapon S, Takahashi K (2008) Land subsidence resulting from groundwater pumping in Bangkok. Soil Mech Found Eng 53(2):16–18

Ohmachi T, Romam ER (eds) (2002) Metro Manila: in search of a sustainable future. University of the Philippines Press, Quezon City, p 338

Porter RC (1996) The economic of water and waste. Avebury, Aldershot, p 125

Reed RR (1978) Colonial Manila. University of California Press, Berkeley, p 70

Selya RM (1995) Taipei, Wiley, New York, p 266

Seoul Museum of History (2006) The maps of Seoul. Seoul Museum of History, Seoul, p 248

Taniguchi M et al (2009) Changes in the reliance on groundwater versus surface water resources in Asian cities. IAHS Publ 330:1–7

Tasaka T (ed) (1998) Big cities in Asia[1]. Nippon Hyoronsha, Bangkok, p 335

Wada K, Nonaka T, Sano T (2006) Issue of water service privatization in Jakarta. Water Sci 290:52–76

Yoshihara K (2003) Economic growth and environment in Taiwan. International Center for the Study of East Asian Development, Working Paper Series Vol. 2003-32, p 20

Yoshikoshi A et al (2009) Hydro-environmental changes and their influence on the subsurface environment in the context of urban development. Sci Total Environ 407:3105–3111

Yoshikoshi A (2010) Urban development and water environment change in Asia. In Taniguchi M (ed) Subsurface environment of Asia: the left-behined global environment problems. Gakuhosha, Tokyo, pp 67–88

Chapter 4
Comparative Analysis on Land Use Distributions and Their Changes in Asian Mega Cities

Akio Yamashita

Abstract Distribution of land use and its chronological changes are the mirrors that directly reflect the present situation and changes in the natural and socioeconomic environments in the region concerned, and they serve as indexes to measure the effects of people's activities on the ground. As such, analysis of land use is the basis for academic fields that approach the relationship between human activities and nature from a spatial perspective, such as geography. The purpose of this study is to establish the versatile method to create land use mesh data from topographic maps including old-edition maps of various countries, and to comparatively analyze spatial characteristics of land use distribution in respective cities in respective periods, and their chronological changes. As a result, I was able to analyze and interpret the land use distribution patterns and changes in the past century in seven cities in Asia. The method of this study is considered to be versatile for preparing the same standard land use maps, targeting a relatively wide range, such as metropolitan areas, and using overseas maps and old-edition maps.

4.1 Introduction

Distribution of land use and its chronological changes are the mirrors that directly reflect the present situation and changes in the natural and socioeconomic environments in the region concerned, and they serve as indexes to measure the effects of people's activities on the ground. As such, analysis of land use is the basis for academic fields that approach the relationship between human activities and nature from a spatial perspective, such as geography.

Most of the megacities in Asia are located on low-laying areas alongside the downstream of large rivers; the population has rapidly increased and these cities

A. Yamashita (✉)
Graduate School of Life and Environmental Sciences,
University of Tsukuba, 1-1-1 Tennodai, Tsukuba, Ibaraki 305-8572, Japan
e-mail: akio@geoenv.tsukuba.ac.jp

M. Taniguchi (ed.), *Groundwater and Subsurface Environments: Human Impacts in Asian Coastal Cities*, DOI 10.1007/978-4-431-53904-9_4, © Springer 2011

have changed significantly in the past period of 50–100 years. Accompanying these changes, problems in water resources, urban heat island effects, and underground environment issues symbolized by land subsidence have occurred. However, the present stage of progress and maturity as a city differs among cities, and as a result, the overt and latent characteristics noted in the above-mentioned urban environment issues also vary.

In this study, therefore, land use mesh maps were made to be used as indexes to compare megacities in Asia on the progress of urbanization and industrialization and accompanying various urban environmental problems. This study examined seven cities in three periods: Tokyo, Osaka, Seoul, Taipei, Manila, Bangkok, and Jakarta. Because these cities are capital cities (or cities equivalent of capital cities) and have different stages of urban development each other, they are selected as the target of this study. By Yoshikoshi (2010), Tokyo and Osaka are developed first, and Seoul and Taipei are behind them. Manila and Jakarta are developed later. The purpose of this study is to establish the versatile method to create land use mesh data from topographic maps including old edition maps of various countries, and to comparatively analyze spatial characteristics of land use distribution in respective cities in respective periods, and their chronological changes. Some previous studies such as a series by Institute for Economic Research, Osaka City University (Tasaka 1998, Miyamoto and Konagaya 1999, Nakanishi et al. 2001, and so on) targeted on some Asian mega cities and detailed their histories of urban development. They described urban development qualitatively, while this study is intended to analyze it quantitatively.

First, I took a general view of the state of preparation of land use mesh data and related existing studies in Japan. In this country, digital data regarding land use are prepared and disclosed mainly by the Ministry of Land, Infrastructure and Transport, and land use can be analyzed relatively easily. Examples include studies by Sugimori and Ohmori (1996) and by Yamashita (2004). However, these existing digital data were on the 1970s to 1990s. Only recently were data on 2000s disclosed, but there is no data on the 1960s or earlier.

On the other hand, as studies that restored the past land use by using the old editions of topographic maps and comparing them with the present state, there are a series of studies by Yukio Himiyama (Himiyama and Wataki 1990, Himiyama et al. 1991, Himiyama and Ota 1993, Himiyama and Motomatsu 1994). In these studies, land use data were prepared in the units of 2 km mesh throughout Japan based on the 1:50,000 topographic maps in Meiji, Taisho and mid-Showa periods (1950s), and comparisons were made with the present. These are very valuable data for learning about the state of Japan back then, but it is estimated that data preparation took vast amounts of time and work. Few studies on changes in land use on such a time scale have followed.

On the other hand, regarding cities overseas, there are sets of data on land use read from the satellite images such as Landsat and MODIS. The oldest satellite images are those from the 1970s, and there are none before that. Himiyama et al. attempted restoration of land use diagrams based on old editions of Chinese topographic maps (Himiyama et al. 1995, 1997, 1998, 1999), but no study has restored land use diagrams on a time scale of 50–100 years regarding the five overseas cities that our study targets. Also, the government agencies of various countries have prepared and disclosed digital data on land use, but they were prepared according to respective countries' original preparation standards and legend divisions, so it is difficult to say that these data can be simply applied directly, considering the key point of this study, which is to compare cities.

4.2   Methodology

In order to compare cities and periods regarding land use, which is the aim of this study, it is necessary to prepare land use mesh maps using the same standards and procedures as much as possible. On this occasion, the first issue is the reduction scale of the topographic maps, which are the base maps. The reduction scales of Japanese topographic maps come in three types: 1:25,000, 1:50,000, and 1:200,000, but foreign topographic maps do not necessarily use the same scales. Regarding this point, the published topographic maps of five cities that I target in this study were checked. As a result, it was found that the maps with the 1:50,000 reduction scale were published for four out of the five cities. As such, I decided to collect topographic maps of seven cities and three periods with this reduction scale, to the extent possible. Regarding the maps of Jakarta in the 2000s, however, only those with the reduction scale of 1:25,000 are published, so I used them as substitutes.

Next, the spatial ranges of cities included in this study were considered. The definition of the term "metropolitan area" varies depending on the measure, and in this study I demarcated the target ranges from the perspective of analyzing land use. Specifically, I targeted the areas that included the entire conurbation (urban districts) from respective city centers, or areas slightly larger than them. Our range on the time axis was based on a scale of about 100 years, but I used maps from the 1920s or 1930s as old maps, considering the following two factors: whether 1:50,000 old-edition topographic maps, which are the base maps, can be prepared according to the almost same periods when they were prepared for all target maps; and the avoidance of overlapping with prior studies by Himiyama et al. And also, the current maps (2000s) and maps of the 1950s or 1960s were prepared for this study.

Thirdly, the mesh size was considered, which is the data unit. Studies by Himiyama et al. targeting all of Japan used a 2 km mesh. This study, however, targets only metropolitan areas, so the mesh size should be smaller than that. Ohbi (2008) used 1/4 subdivided mesh (250 m mesh) targeting northern part of Kanto Region, but use of the same size in this study is prospected to require a huge amount of time and labor for data preparation. In other words, it is not good to only get rough ideas about distribution and changes of land use as data for the spatial scale of the areas studied. On the other hand, data preparation becomes too cumbersome if excessive details are pursued. Considering these factors, this study decided to adopt the 1/2 subdivided mesh (500 m mesh).

The last consideration was on how the land use items should be classified in legends. I opted for less divisions without setting up detailed classified items, with reference to 11 divisions of the "land use mesh of digital national land information," which is the existing land use mesh data in Japan, as well as 16 divisions of "detailed digital information." The reason is as follows: in order to comparatively analyze land use in the seven cities in three periods that are targeted in this study, I have to prepare data of respective cities and periods with the same classification items; and for that purpose, I need to narrow classification items down to those that can be read from the map of every city and period. On the other hand, this study aims to relatively consider spatial-temporal features of urbanization and industrialization in respective cities based on the analysis of land use distribution. To describe urbanization from the perspective of land use, it is essential to clarify the divisions of natural green land, agricultural land, open space and urban area. Also, because the seven cities in Asia that are targeted in this study are located in places rich in water along big rivers, it is also interesting to find distribution of water areas inside city districts. Considering these factors, I adopted the following nine land use items in this study: (1) "forest" (needle-leaved trees, broad-leaved trees, bamboo grove), (2) "grassland, wasteland" (including park, artificial green land and golf course), (3) "rice field," (4) "other agricultural lands" (field, orchard, pasture), (5) "industrial site," (6) "residential area" (urban land use other than industrial site), (7) "water area, wetland," (8) "others" (developed land, unused land, etc.), and (9) "ocean."

After setting up unified standards as mentioned above, the actual data preparation was started. The procedure is to import the base maps using a scanner, and enter positional information as defined image data in GIS. Then, it was overlapped with the separately prepared 1/2 subdivided empty mesh file, and what stood out most in terms of area in each mesh based on the above nine land use items was visually checked and entered as attribute information in the mesh file (as for detailed preparation procedure, refer to Yamashita et al. 2008, 2009). Regarding overseas cities, however, many places were very difficult to read even with this nine-item classification, due to differences in the accuracy of the base map and the definitions of map symbols. Reliability of these data undoubtedly needs to be verified in the future. In the stage of this report, (1) "forest" and (2) "grassland, wasteland" were combined as "natural green land"; (3) "rice field" and (4) "other agricultural lands" were classified as "agricultural land (productive green land)"; (5) "industrial site" and (6) "residential area" were classified as "urban area"; and the rest was classified as "others." The three divisions other than "others," were analyzed the spatial distribution characteristics and time changes.

4.3   Result

4.3.1   Tokyo

Figure 4.1 shows land use mesh maps of three periods in Tokyo Metropolitan Area. The range of the urban area in Tokyo in 1927 was a radius of about 10 km. On the plain area along the large river at the outer edge of the urban area,

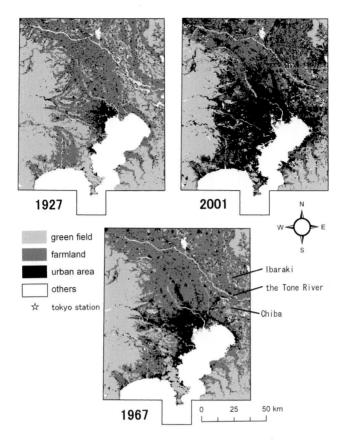

Fig. 4.1 Land use mesh maps of three periods in Tokyo

especially from the northwest to northeast part, agricultural land was dominant. In 1967, the urban area had expanded along major traffic networks. Agricultural land significantly expanded in the southern part of Ibaraki prefecture and the northern part of Chiba prefecture along the Tone River. In 2001, the urban area expanded to a radius of about 40 km, and many urban areas acting as centers for the suburbs are seen even farther away. Many of these urban areas used to be agricultural land.

Next, I see the changes in land use ratios according to distance zones of 10 km each from Tokyo Station as the representative point of the city center, based on the GIS buffer analysis (Fig. 4.2). To calculate the land use ratio, "ocean" mesh and mesh without data are excluded (the same with other cities). In the Tokyo area, natural green land decreased and agricultural land increased in regions farther than 30 km from the city center, from the 1920s to 1960s. From the 1960s to nowadays, agricultural land has significantly decreased and urbanization has progressed.

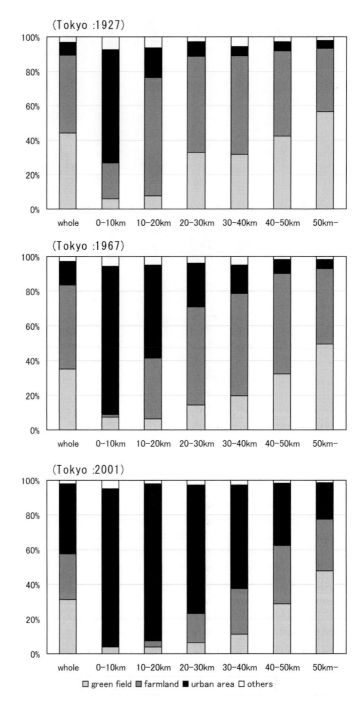

Fig. 4.2 Land use ratios by distance zones of every 10 km in Tokyo

4.3.2   Osaka

Figure 4.3 shows the land use mesh maps of three periods in Osaka Metropolitan Area. In the urban area in the Osaka area in 1927, there were three centers: Osaka city, Kyoto city and Kobe city, and each was about 5 km wide in radius. Agricultural land accounted for a considerable part around Osaka, Nara and Kyoto cities, and the lakeside of Lake Biwa at the time. The majority of the agricultural land was converted to urban area by 2001. On the other hand, the areas that used to be natural green land were not converted to agricultural land or urban area due to topographical restrictions, and they have remained as green land.

Next, Osaka Station was adopted as the representative point in the city center of the Osaka area and changes in land use ratios were checked according to distance zones of every 10 km from there (Fig. 4.4). In the Osaka area, a considerable area of agricultural land was seen even within the 10 km sphere from the city center in the 1920s, but it has sharply decreased in both the inner-city districts and suburban districts. On the other hand, the ratio of natural green land is characterized by no significant change in any distance zone from 1920s up to now.

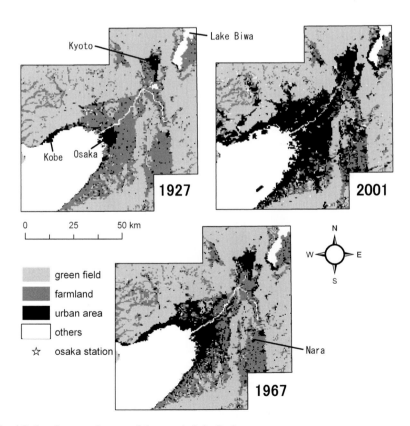

Fig. 4.3 Land use mesh maps of three periods in Osaka

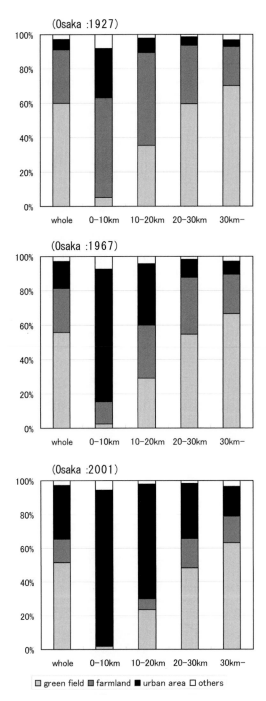

Fig. 4.4 Land use ratios by distance zones of every 10 km in Osaka

4.3.3   Seoul

Figure 4.5 shows the land use mesh maps of three periods in Seoul. The urban area in Seoul in the 1920s or 1930s was seen only in the limited range in the northern part of the Han River, but in the 1960s the area expanded to the southeast part on the opposite side of the river. Agricultural land was distributed linearly between mountainous areas. In the 2000s urban areas were widely distributed on both banks of the Han River, and also in Inchon in the west and Anyang and Suwon in the south, urban areas acting as suburban centers were formed.

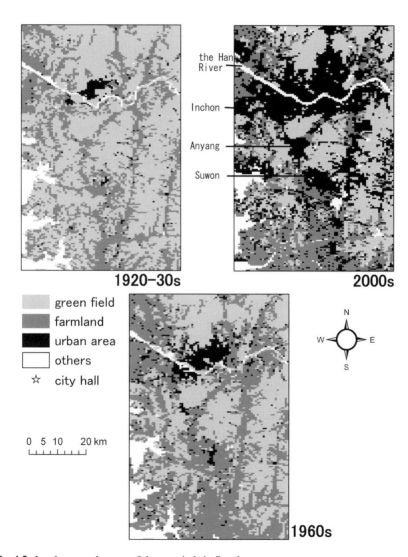

Fig. 4.5 Land use mesh maps of three periods in Seoul

Next, regarding the features according to distance zones from the city center, Seoul city hall was chosen as the representative point of the center of Seoul city. The metropolitan area ranges of the four overseas cities are smaller compared with Tokyo and Osaka, so I tallied their land use ratios using distance zones of every 5 km. According to Fig. 4.6, the urban area in Seoul was within the range of the 5 km radius in the 1920s or 1930s, and it was within the range of the 10 km radius even in the 1960s. Natural green land accounted for 60% or more in the 1920s or 1930s, but it significantly decreased in all distance zones by the 2000s. Green land was converted mainly to urban areas within the 10 km sphere and to agricultural land farther than 10 km by the 1960s. The converted agricultural land has been further converted to urban areas in recent years.

4.3.4 Taipei

Figure 4.7 shows the land use mesh maps of three periods in Taipei. The urban area in Taipei in the 1920s or 1930s was very small in scale compared with other cities, and it was surrounded by agricultural land. By the 2000s, areas that used to be agricultural land were converted to urban areas, while natural green land sites remained relatively as they were due to topographical restrictions.

Figure 4.8 shows the calculation of land use ratios according to distance zones from Taipei Station as the center of the inner city. This figure also indicates that the urban area in Taipei was also within the radius of about 10 km in the 2000s, and areas farther than 10 km are mostly occupied by natural green land.

4.3.5 Manila

Figure 4.9 shows the land use mesh maps of three periods in Manila, but in the 1930s, the accuracy of the base map was low; many map symbols that were not included in the legend were used and map symbols varied depending on maps. These factors made it very difficult to read land use. Figure 4.9 also shows a clear discontinuity of land use in the borders of maps, indicating the need to verify data accuracy.

According to Figs. 4.9 and 4.10, the latter of which tallied land use ratios according to distance zones from Manila city hall as the center of the city, the urban area in Manila in the 1960s was within a radius of about 10 km, and agricultural land was dominant in the northwest and southwest parts close to the ocean outside the urban area. By the 2000s, the urban area expanded to a radius of about 20 km in a radial pattern, and urbanization of the area near Lake Laguna in the south is particularly noteworthy.

4.3.6 Bangkok

Figure 4.11 shows the land use mesh maps of three periods in Bangkok. The urban area in Bangkok in the 1950s was within a radius of 5 km, but the majority of the

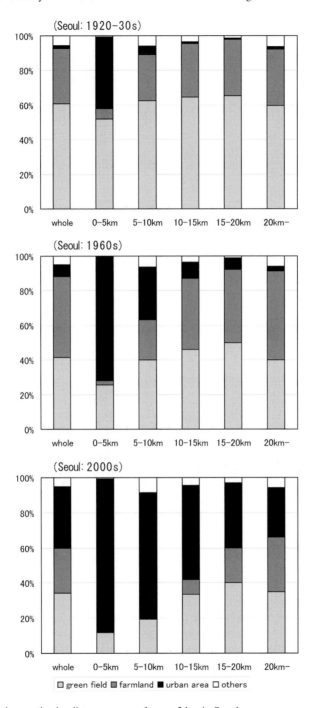

Fig. 4.6 Land use ratios by distance zones of every 5 km in Seoul

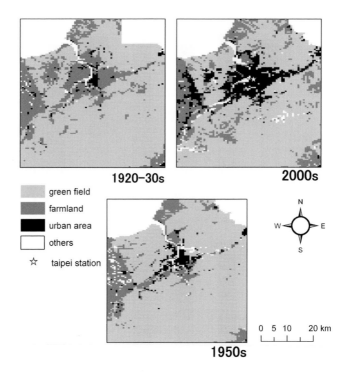

Fig. 4.7 Land use mesh maps of three periods in Taipei

surroundings had already been converted to agricultural land, and almost no natural green land was seen. In the 2000s the urban area expanded in a radial pattern just like in Manila.

Regarding the land use ratios according to distance zones from the royal palace (Fig. 4.12), the urban area in Bangkok in the 2000s seems to have expanded to a radius of about 20 km, especially to the eastern side of the Chao Phraya River. In areas farther than 20 km, the land use ratio of "others" is high, and this is because fish ponds are distributed in a considerably wide range along the coast.

4.3.7 Jakarta

Figure 4.13 shows the land use mesh maps of three periods in Jakarta. In Jakarta in the 1930s, collective urban areas were formed around Kota, which was an old inner-city district since the time of governance by Holland, and around Gambir, which is the current center; in addition, small urban areas as centers of settlements were scattered. And, agricultural land was distributed around the centers of these settlements. Even in the 1960s, this tendency did not show significant change, and inner-city districts only slightly expanded. However, a considerable area of natural green land in the south was converted to agricultural land. By the 2000s, urban area widely expanded.

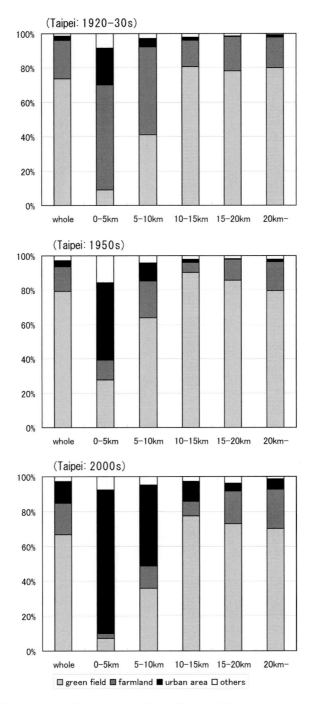

Fig. 4.8 Land use ratios by distance zones of every 5 km in Taipei

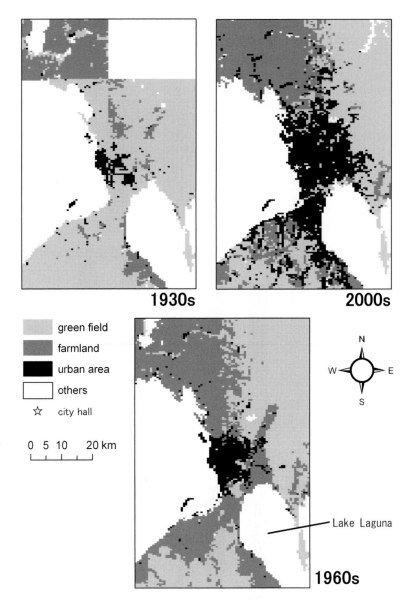

Fig. 4.9 Land use mesh maps of three periods in Manila

The land use ratios show no tendency of sharp changes according to distance zones from Kota Station (Fig. 4.14) as they do in other cities, and in the 1930s, the urban area ratio was about 20% in the 10–15 km zone. The situation was similar even in the 1960s, and it is characterized by the gradual change in the land use ratio according to the distance from the city center; the ratio of natural green land is low as in Bangkok.

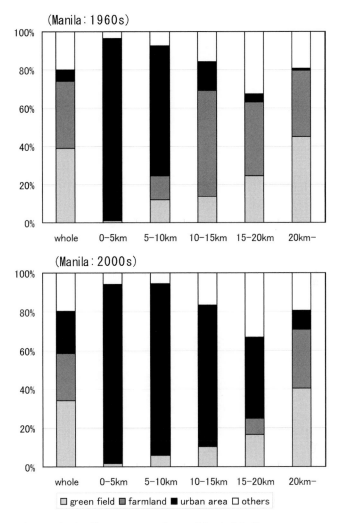

Fig. 4.10 Land use ratios by distance zones of every 5 km in Manila

4.4   Discussion

First, the following is a summary of the analysis results on the Tokyo and Osaka areas. In the plain area in eastward and northward directions from the inner city in particular, along the outer edge of the Tokyo area, two stages of land use changes have occurred: from natural green land to agricultural land, and then to urban area. In the Osaka area and in the westward direction in the Tokyo area, a change from agricultural land to urban area was seen from the 1920s to 2000s, but natural green lands in mountainous regions remained unchanged.

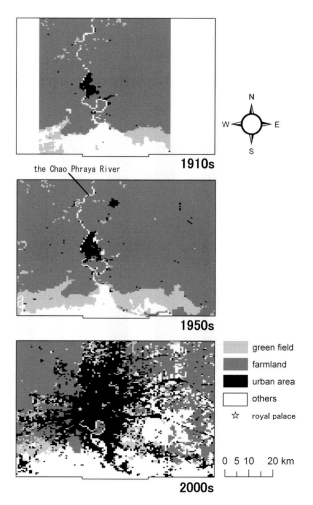

the Chao Phraya River

1910s

1950s

green field

farmland

urban area

others

☆   royal palace

0 5 10   20 km

2000s

Fig. 4.11 Land use mesh maps of three periods in Bangkok

Next, comparison of spatial expansion of the present urban areas in seven cities indicates that the urban area is the largest in Tokyo, followed by Osaka. As for overseas cities, the urban area is large in Seoul, Bangkok and Jakarta, and it expands into the range over 20 km in radius from each city center. Then, Manila follows. Taipei's urban area is the smallest with a radius of about 10 km. The reasons for this difference of urban expansion are mainly topographical constraint and age of modernization. Spatial expansion of urban area is restricted by Lake Laguna in Manila. As Taipei is located in the basin and surrounded with mountainous area, the urban area is hard to expand. Jakarta and Manila were modernized later because they were colonized until the end of World War II. Comparison between Bangkok and Seoul reveals that in Bangkok, the urban area expands concentrically from the royal palace district, which is the city center, while in Seoul, urban cores are seen

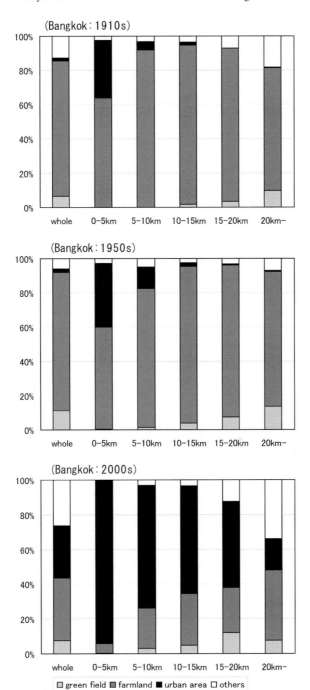

Fig. 4.12 Land use ratios by distance zones of every 5 km in Bangkok

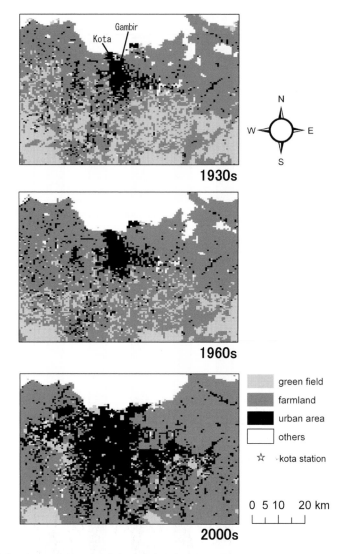

Fig. 4.13 Land use mesh maps of three periods in Jakarta

also in the west and south in addition to the region centered on the inner-city district
along the Han River, presenting a multi-nucleus city structure. In Osaka, Seoul, and
Taipei, the urban areas are surrounded by natural green land. On the other hand, in
Bangkok and Jakarta, large agricultural land surrounds urban areas, and there are
almost no natural green land such as forests and grassland in the target areas. This
indicates that in Bangkok and Jakarta, the natural environment such as forests and
grassland has artificially been altered over a wide range from early on. These relative characteristics also seem to come from topographical conditions.

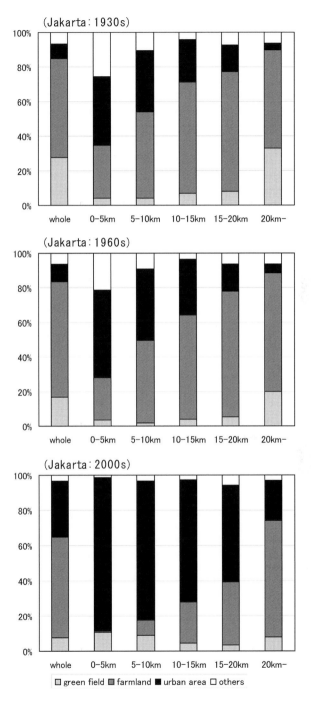

Fig. 4.14 Land use ratios by distance zones of every 5 km in Jakarta

4.5   Conclusion

This study used the 1/2 subdivided mesh size and prepared land use mesh maps using a simple and efficient method of visually reading topographical map images. As a result, this study could analyze and interpret the land use distribution patterns and changes in the past century in seven cities in Asia. The method of this study is considered to be versatile for preparing the same standard land use maps, targeting a relatively wide range, such as metropolitan areas, and using overseas maps and old-edition maps. In other words, this method is suitable for a comparative study that compares the past and present, and overseas cities and cities in Japan. As future works, I will conduct relation analysis through simultaneous checking of data on the natural environment, including landform, geology, water quality, and so on. Subsequently, I will consider the causes and countermeasures on urban environmental problems such as subsurface problems including groundwater quantity and land subsidence, and a problem on balance of urban water supply and demand.

Acknowledgements This research was financially supported by the project "Human Impacts on Urban Subsurface Environment" (Project Leader: Makoto Taniguchi), Research Institute for Humanity and Nature (RIHN). I express my gratitude to Prof. Akihisa Yoshikoshi and other members of this project for their cooperation to collect base maps, and to Dr. Chun-Lin Kuo (National Dong Hua Univ., Taiwan), Yayumi Abe (EnVision Conservation Office), Kiyoshi Takaoku (Sancoh Corporation) and students of Rakuno Gakuen University for their works to create land use mesh data.

References

Himiyama Y, Motomatsu H (1994) Land use change in the Tohoku District since circa 1910. Rep Taisetsuzan Inst Sci 29:1–16 (in Japanese with English abstract)
Himiyama Y, Ota N (1993) Land use change of Hokkaido since circa 1920. Rep Taisetsuzan Inst Sci 28:1–13 (in Japanese with English abstract)
Himiyama Y, Wataki N (1990) Reconstruction of land use of Hokkaido in the Taisho Period. Rep Taisetsuzan Inst Sci 25:25–34 (in Japanese with English abstract)
Himiyama Y, Iwagami M, Inoue E (1991) Reconstruction of land use of Japan circa 1900–1920. Rep Taisetsuzan Inst Sci 26:55–63 (in Japanese with English abstract)
Himiyama Y, Ito H, Kikuchi T, Honma T (1995) Land use in North-East China in the 1930s. Rep Taisetsuzan Inst Sci 30:25–35 (in Japanese with English abstract)
Himiyama Y, Fujisawa M, Miyakoshi T (1997) Land use in North-East China circa 1980. Rep Taisetsuzan Inst Sci 31:13–23 (in Japanese with English abstract)
Himiyama Y, Suzuki S, Hayakawa A (1998) Reconstruction of land use in the Southern North China Plain in the early 20th century. Rep Taisetsuzan Inst Sci 32:13–22 (in Japanese with English abstract)
Himiyama Y, Iwamoto S, Watanabe E (1999) Land use in the Southern North China Plain circa 1910–1980. Rep Taisetsuzan Inst Sci 33:9–18 (in Japanese with English abstract)
Miyamoto K, Konagaya K (eds) (1999) Asian mega city 2 -Jakarta-. Nihonhyoronsha, Tokyo (in Japanese)
Nakanishi T, Kodama T, Niitsu K (eds) (2001) Asian mega city 4 -Manila-. Nihonhyoronsha, Tokyo (in Japanese)

Ohbi K (2008) Landscape characteristics from the viewpoint of river systems in watersheds: a case study of the Nakagawa, Kasumigaura, Kinugawa, and Kokaigawa Watersheds in the North Kanto Region, Japan. J Geogr 117:534–552 (in Japanese with English abstract)

Sugimori H, Ohmori H (1996) Study of explicating landscape dynamics from land use in the middle and lower basin of the River Tamagawa, Tokyo Metropolitan Area. Theory Appl GIS 4(2):51–62 (in Japanese with English abstract)

Tasaka T (ed) (1998) Asian mega city 1 -Bangkok-. Nihonhyoronsha, Tokyo (in Japanese)

Yamashita A (2004) Land use characteristics on watershed scale and their regional differences in Japan. Pap Proc Geogr Inf Syst Assoc 13:79–82 (in Japanese with English abstract)

Yamashita A, Abe Y, Takaoku K (2008) The creation of land use mesh data based on topographic maps in Asian mega cities. Pap Proc Geogr Inf Syst Assoc 17:205–208 (in Japanese with English abstract)

Yamashita A, Abe Y, Takaoku K (2009) The creation of land use mesh data on old topographic maps and analysis of land use changes in Tokyo and Osaka Metropolitan Areas. Pap Proc Geogr Inf Syst Assoc 18:529–534 (in Japanese with English abstract)

Yoshikoshi A (2010) Urban development and change of water environment in Asia. In: Taniguchi M (ed) Subsurface environment in Asia. Gakuhosha, Tokyo (in Japanese)

Part II
Groundwater Degradation and Resources Management

Chapter 5
Monitoring Groundwater Variations Using Precise Gravimetry on Land and from Space

Yoichi Fukuda

Abstract In order to establish a new technique for monitoring the groundwater variations, we investigated the applicability of precise in-situ gravity measurements and the GRACE (Gravity Recovery and Climate Experiment) satellite gravity data. A new scheme for in-situ measurements that combines absolute and relative gravity measurements as well as GPS measurements has been proposed to monitor groundwater variation and associated land subsidence as well. For this purpose, we introduced a portable type absolute gravimeter (Micro-G LaCoste Inc. A10) which can be used for field surveys. We conducted several test surveys and confirmed that the gravimeter can achieve a 10 μgal (100 nm/s^2) or better accuracy in the field surveys. GRACE is providing extremely high precision gravity field data from space. These data are precise enough to reveal the gravity changes due to large scale groundwater variations. Using the GRACE data, we estimated terrestrial water storage (TWS) variations in the Indochina Peninsula. The results showed good agreements with Soil-Vegetation-Atmosphere Transfer Scheme (SVATS) models basically. The agreements can be improved by tuning the model parameters such as current velocity of river flow. It means that the GRACE TWS can be used as a constraining condition of the models. We also detected the mass trends in the Indochina Peninsula and the gravity changes due to the 2006 drought in Australia. This suggested that GRACE data should be applicable for monitoring secular or long-term groundwater variations at the continental scale as well. Finally, as a future prospection, we discuss the role of hydrological models which connect in-situ and satellite observations.

Y. Fukuda (✉)
Department of Geophysics, Kyoto University, Kitashirakawa Oiwake-cho, Sakyo-ku,
Kyoto 606-8502, Japan
e-mail: fukuda@kugi.kyoto-u.ac.jp

M. Taniguchi (ed.), *Groundwater and Subsurface Environments: Human Impacts in Asian Coastal Cities*, DOI 10.1007/978-4-431-53904-9_5, © Springer 2011

5.1   Introduction

Geodesy in the twenty-first century is evolving into a cross-disciplinary science and engineering discipline. The strength of blending accurate instruments on space as well as on land enables us to address contemporary problems such as monitoring of Earth's environmental issues. Water resources transform between groundwater and surface water in many cities depending on the urban development stage. As a part of the Research Institute for Humanity and Nature (RIHN) project 2.4 Human Impacts on Urban Subsurface Environments (HIUSE) (Taniguchi et al. 2008), we have conducted the researches to evaluate groundwater flow systems in and around developing cities.

When precisely measuring gravity on land, groundwater variations were one of the largest noise sources so far, especially for high precision gravity measurements using absolute gravimeters. The effects, if appropriately analyzed, can thus give us important information about the hydrological characteristics in the area concerned. In fact, water mass changes can be directly measured only by gravimeters, thus the technique is expected to contribute to water resource management.

Another issue closely related with the groundwater changes is land subsidence. Excessive groundwater extraction leads to the deepening of piezometric head. This may cause the sinking of the upper layer, i.e., land subsidence, and in coastal areas, seawater intrusion as well. Although the main cause of the land subsidence should be the groundwater extraction, there are some other factors, namely, the load of constructed buildings, natural consolidation of soil and geotectonic processes. The amount of land subsidence can be measured by repeated leveling measurements or GPS surveys (e.g., Abidin et al. 2008). However we need additional information, such as groundwater level and/or gravity changes, to identify the cause of the land subsidence. Again, the combined survey of gravity, GPS and groundwater level measurements should be a powerful tool for the studies of land subsidence as well.

In Chap. 2, we described the examples of the surface gravity measurements, in particular about the field absolute gravity measurements using the Micro-g LaCoste Inc. (MGL) A10 gravimeter. We described the technical aspects of the gravimeter and referred to some examples of the practical surveys.

GRACE (Gravity Recovery and Climate Experiment), on the other hand, is providing extremely high precision gravity field data from space. The GRACE data have been employed for various hydrological applications (e.g., Tapley et al. 2004b; Chen et al. 2005). Although its spatial resolution is not enough to reveal the urban scale variations, GRACE provides the information about a regional scale terrestrial water storage (TWS) variations which is indispensable for the studies of urban scale groundwater variations. Therefore we firstly estimated the seasonal mass variations in four major basins of the Indochina Peninsula, and compared them with a TWS model. Furthermore, using the most updated GRACE data, we have revealed not only seasonal variations but also a secular trend of the mass variations in the Chao Phraya river basin. We also succeeded to detect the mass changes associated with the 2006 Australian drought.

In Chap. 3, we first described the GRACE mission and the outline of the data processing. And then, we referred to the studies in the Indochina Peninsula and the Australian drought as well.

In Chap. 4, we summarized the gravity methods for monitoring groundwater variations and we discussed the future prospection to combine the satellite and in-situ observations from the viewpoints of the multi-scale hydrological models.

5.2   In Situ Measurements

5.2.1   Role of Precise Gravity Measurements

Local hydrological variations crucially affect precise gravity measurements, for instance, relative gravity observations (e.g., Lambert and Beaumont 1977), super-conducting gravity observations (e.g., Abe et al. 2006), and absolute gravity measurements (e.g., Bower and Courtier 1998). However, practical application of the gravity method for hydrological studies is very limited. One of the successful applications of this kind of study may be for geothermal power stations (Allis and Hunt 1986). It is necessary to monitor the mass balance in geothermal reservoirs to produce geothermal fluid (steam and hot water) over a long period. For instance, Nishijima et al. (2007) conducted repeated gravity measurements at Takigami geothermal field located in central Kyushu, Japan. In general, the expected gravity signals at the geothermal field are significant, reaching several tens of μgals (10^{-8} m/s^2) or more. Nishijima et al. (2007) employed Scintrex CG-3 and CG-3M gravimeters for measuring precise gravity changes around the Takigami geothermal power station and attained an observation accuracy of ±10 μgals, which was sufficiently accurate to detect the signals. Another example is repeated gravity measurements using LaCoste and Romberg gravimeters at Bulalo geothermal power station, Philippines (Nordquist et al. 2004). They used the gravity changes for constraints in a simulation model in order to reflect the natural recharge.

Not many studies have been conducted to monitor groundwater variations or to investigate hydrologic problems, primarily because the expected signals are small compared with the geothermal applications. However, the basic principle of the hydrological application is rather simple; the gravity changes due to groundwater mass movements can be measured by means of precise gravimeters. An infinite water table of 1-m thickness causes about a 40-μgal gravity change. Thus, an accuracy of 10 μgals or better is required for the hydrologic problems. It is not easy to achieve an accuracy of 10 μgals by means of a spring-type relative gravimeter, for instance Schintrex or LaCoste & Romberg gravimeters. We therefore proposed a new method to combine absolute gravity measurements and relative gravity measurements. Figure 5.1 shows the configuration of proposed hybrid gravity measurements with relative and absolute gravimeters combined. For the measurements at the control points, we employ a portable absolute gravimeter A-10 of MGL

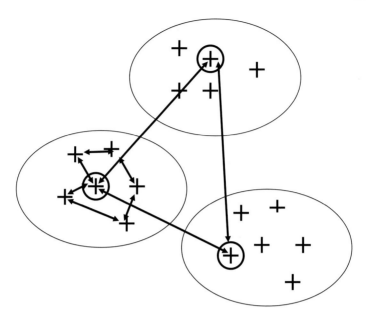

Fig. 5.1 A schematic illustration of the hybrid gravity measurements. The *cross-shape marks* show the gravity points for relative gravity measurements and the *cross-shapes with a circle* show the control points where the absolute gravity measurements are conducted

(Micro-g LaCoste Inc.) (2008), while relative gravimeters of superior portability are employed for the measurements at most points around the control points. One sequence of relative measurements should be completed within a few hours or less to minimize the drift errors. The hybrid gravity measurements with both absolute and relative gravimeters can strike a balance between accuracy and efficiency of the measurements.

5.2.2 A-10 Absolute Gravimeter

As described above, employment of a portable absolute gravimeter is essentially important for the hybrid gravity measurements. We introduced the A-10 portable absolute gravimeter for this purpose. As for the absolute gravity measurements, FG-5 of MGL is well known. FG-5 is a high precision absolute gravimeter with a 2-μgal-accuracy for laboratory use. Because of this high precision, FG-5 is most commonly employed for the absolute gravity studies. A-10 is a modified version of the FG-5 for outdoor field surveys. It is much smaller than FG-5 and can be operated with 12VDC power. As the trade-off, the nominal accuracy of A-10 is 10 μgal (measurement precision: ±5μgal). It is slightly worse than that of FG-5 but satisfactory for many of the hydrological studies.

5.2.2.1   Overview of A10

In this section, we briefly describe an overview of the A10 absolute gravimeter. The detailed descriptions are found in the User's Manual [MGL (Micro-g LaCoste Inc.) 2008] which is downloadable from MGL web site.

Figure 5.2 shows the principle of the absolute gravity measurements by the A10 gravimeter. The A10 is a ballistic absolute gravimeter which measures the gravity as the vertical acceleration of a dropping test mass (corner cube). The test mass is freely falling in a vacuum chamber and its dropping distances and times are measured with a laser interferometer and a rubidium atomic clock, respectively. The interferometer basically consists of a beam splitter (a half mirror) and two corner cube retro-reflectors; one is the dropper corner cube and the other is the reference corner cube. The laser beam led by laser fiber is split by the beam splitter into the test and reference beams. The test beam is reflected by the dropper corner cube and the reference corner cube and finally recombined with the reference beam. The reference corner cube is supported by the super-spring, which is a kind of long period seismometer, to isolate land vibration noises. The distance change of the dropping corner cube causes interference fringes which are detected by an optical detector (photo-diode). The fringes counted and timed with the atomic clock provide the data of precise time and distance pairs. The vertical acceleration is calculated by fitting these data to a parabolic trajectory.

The A10 can automatically lift up and drop the test mass every 1 s. We usually combined 100 drops of measurements into a "set," and conducted ten sets of measurements to obtain an absolute gravity values. According to the MGL web site, the precision of the A10 at a quiet site is 50 µgal/sqrt (Hz). Therefore the ten sets of

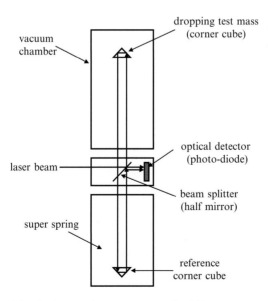

Fig. 5.2 Principle of the absolute gravity measurements by A10

measurements (1,000 drops) can attain almost 1 µgal precision. On the other hand, the accuracy of A10 is restricted by numerous factors including the tides and other geophysical corrections. MGL announced that the accuracy of A10, which means observed agreement between A10 instruments, is 10 µgal. Schemerge and Francis (2006) investigated and discussed about the precision and the accuracy of an A10 (A10-008: serial number 8) in details. They confirmed A10-008 showed better performance than the specifications. As described in the next section, we also confirmed the A10 cleared the specifications and frequently showed better performances in good condition sites.

5.2.2.2   Test Measurements Using A10-017

In order to carry out the RIHN project 2.4, A10-017 had been newly introduced in December 2007. Since then, we have conducted several test measurements both indoors and in the fields. Figure 5.3 shows a photo of A10-017.

The first test was the comparison with another absolute gravimeter. For this purpose, we conducted a measurement with A10-017 at an absolute gravity point in Kyoto University and compared the observed gravity values with the one by FG5-210. The result showed that the agreement within a few µgals. We have occasionally conducted the measurements at the same gravity point and confirmed that the observed values were within 10 µgals. It is known that there exist seasonal gravity variations within 10 µgals in Kyoto due to local groundwater variations (Fukuda et al. 2004). Therefore the observed gravity variations may contain a part of those signals, although we could not confirm that due to the lack of groundwater level data.

Similar repeated measurements have been conducted more often at the gravity point in Fukuoka (Kyushu University) because A10-017 is usually maintained therein.

Fig. 5.3 A photo of A10-017

Figure 5.4 shows all the measured values at the point. There are three groundwater monitoring wells near the points. The groundwater levels observed at the wells are also shown in Fig. 5.4. It can be seen that the gravity values showed a good correlation with the groundwater levels. Considering the gravity signals due to the groundwater variations, the repeatability of A10-017 at the point is better than 10 μgals.

We also conducted several test measurements in the field not only to confirm the accuracy but also to investigate the practical and efficient measurement methods for field surveys. Among them, an interesting result has been obtained through the measurements in the Takigami geothermal field. In a geothermal plant, geothermal fluid (high temp water and vapor) is pumped up from production wells, used for the power generation, and after usage, water is finally retuned into injection wells. This cycle of the geothermal fluid was stopped in April 2008 for a regular maintenance at the Takigami geothermal plant. We thought it was a good opportunity as a field test of the A10 because associated gravity changes should be observed. And we conducted repeated gravity measurements by A10-017 before and after the maintenance. Figure 5.5 shows the observed gravity changes at both the production zone and the injection zone. As we expected, the stop of the production caused a slight gravity increase in the production zone and rather clear gravity decreases in the injection zone. After the restart of the production, the gravity values in both regions were gradually returned to the steady state. We think the gravity signals observed (10–20 μgals) were so small to be hardly detected by a relative gravimeter. Therefore the result proved the efficiency of the A10 measurements. The detailed discussion about the test is found in Sofyan (2009).

It is true that using A-10 is still challenging, especially for groundwater monitoring. However, there is no doubt that the absolute gravity measurements are much superior to the relative measurements. We therefore expect that the A-10 gravimeter will be extensively used for various purposes of the field gravity surveys.

5.2.3   Configuration of the Combined Survey for Groundwater Monitoring

While the mass changes due to groundwater variations should change the gravity value, it is also changed by vertical land movements. The rate of gravity change versus height change depends on the mechanism of height change and associated mass (density) changes. If we assume the mass movement or the density change $\delta\rho$ of underground material (soil or sedimentary layers) and groundwater level changes (δg_w) causes gravity change (δg) as well as height change (δh), then δg can be expressed by the following equation;

$$\delta g = (-0.3086 + 2\pi\delta\rho G)\delta h + 2\pi G \, P_e\delta g_w \qquad (5.1)$$

where G is the Newton's gravitational constant and P_e is effective porosity.

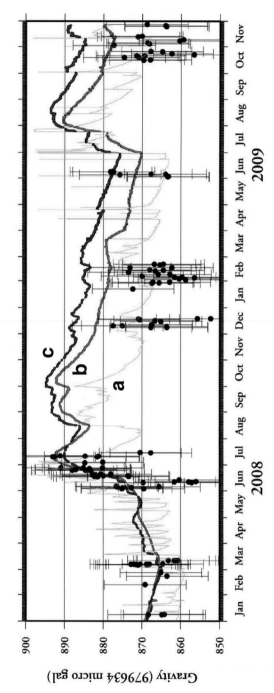

Fig. 5.4 Absolute gravity measurements by A10 in Kyushu University. (**a**), (**b**) and (**c**) show the groundwater variations at the nearby observation wells. Scales of the groundwater variations are converted to "μgals" according to the factors obtained by liner regression analyses. The factors are 4.0, 9.1and 9.8 μgal/m for (**a**), (**b**) and (**c**), respectively

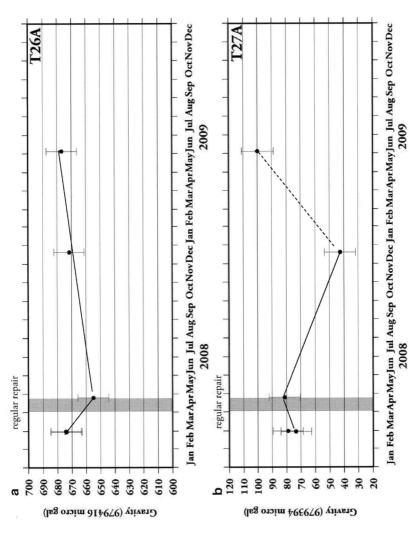

Fig. 5.5 Absolute gravity measurements in Takigami geothermal plant. (**a**) Injection Zone. (**b**) Production Zone

If groundwater level changes of δg_w causes height change of δh, but no density change is associated, i.e., no compaction occurred, then a 1 m vertical movement causes about a 0.3 mgals gravity changes. Therefore height changes at the gravity points should be measured with an accuracy of a few centimeters to ensure the equivalent accuracy of a 10 μgals (0.01 mgals) gravity change.

For monitoring height changes, the leveling is the most accurate method. However it is costly and very time-consuming. Therefore we employed GPS measurements for the purpose. Currently GPS measurements have been employed for many purposes; not only for horizontal positioning but also for height determinations. According to the manual of a commercial GPS receiver (e.g., Trimble 5700), the precision of vertical measurement by static or rapid static survey is ±5 mm + 1 ppm. Moreover many studies have already showed that it is not so difficult to attain 1 cm accuracy in the GPS height measurements.

According to these considerations, we propose a combined method of precise in-situ gravity measurements, GPS measurements and groundwater level measurements for monitoring groundwater variations and associated land subsidence. Figure 5.6 shows the configuration of the combined measurements. At a control point of the hybrid measurements, e.g., a cross-shape with a circle in Fig. 5.1, absolute gravity measurements and GPS measurements should be conducted. Note that gravity

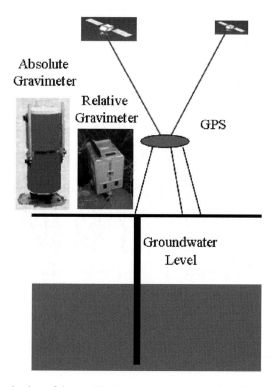

Fig. 5.6 A schematic view of the combined measurements to monitor the groundwater change

and GPS measurements should be occupied at exactly the same marks to ensure the accuracy of a few cm in height or 10 μgals of gravity, whereas groundwater level is not necessarily measured at the same point as long as the measurements well represents the neighboring groundwater level.

5.2.4   Surveys in Jakarta, Indonesia

Jakarta is the capital city of Indonesia and is the largest city in Southeast Asia. It has a population of more than ten million, covering an area of about 650 km². It is located on the lowland of the northern coast of the West Java Province. The area is relatively flat, with topographical slopes ranging between 0° and 2° in the northern and central parts, and between 0° and 5° in the southern part in which the altitude is about 50 m above sea level. There are about 13 natural and artificial rivers which form the main drainage system of Jakarta. We selected Jakarta as a target city because excess groundwater pumping and the resulting land subsidence as well is still going on.

The land subsidence in Jakarta was recognized in 1926. The repeated leveling measurements were conducted in the northern part of Jakarta (e.g., Schepers 1926; Suharto 1971), and the Local Mines Agency reported the cumlative subsidence of 20–200 cm over the period of 1982–1997. The GPS surveys started in 1990s (Abidin et al. 2008) also showed the rate of more than 10 cm/year. Figure 5.7 shows the land movements observed by the GPS surveys from December 2002 to September 2005.

As shown in Fig. 5.7, a good GPS network with more than 20 GPS points has already been established. Therefore we try to make use of these points for the gravity

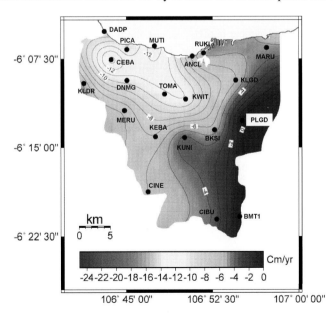

Fig. 5.7 Land movements from December 2002 to September 2005 in Jakarta observed by GPS

measurements, in particular for the absolute measurements, as long as possible. Nevertheless absolute measurements could not be conducted at some GPS points because of mainly logistic reasons. At those points, relative measurements which tied to the absolute point were conducted.

Concerning the groundwater level, there are about 20 observation wells currently operating. As mentioned above, their locations are not necessarily the same as the gravity points. Thus we established some new gravity points near some of the wells. Since the main target of the survey was to detect secular gravity changes associated with groundwater changes and land subsidence, the measurements were conducted at the same season of the year to avoid seasonal variations. Practically it would be good to conduct the measurement in dry season (between July and September) because of the efficiency of the surveys. Also we can reduce several errors due to localized heavy rainfall.

We conducted the first gravity survey in August 2008 and the second survey in July 2009. Referring to the results obtained by GPS surveys conducted so far, the gravity points were selected in the areas with large subsidence. We also selected some point in relatively stable areas for the future references.

Figure 5.8 shows the gravity points. The gravity points marked LIPI and KUNI locate at the stable areas and others located at the areas with large subsidence. Figure 5.9 shows a photo of the A10-017 at KUNI. The A10 gravimeter can be transported by a tailgate mini-van and operated by 12 VDC batteries. The photo shows that it was installed on the GPS point (Bench mark).

Although we had acquired the know-hows for operating the A10 through the test measurements in Japan, it was actually a rather tough work to conduct the measurements in high temperature and humid noisy urban circumstances. In particular the measurements in high temperature caused a problem in vacuum. In order to keep the inside of the dropping chamber in high vacuum, the A10 installs an ion vacuum pump. However the efficiency of the ion pump decreases in high temperature (about 40°C). It occurred that the ion pump of A10-017 could not keep the vacuum enough in the 2008 surveys. Mainly due to the vacuum problem, we could not get enough good

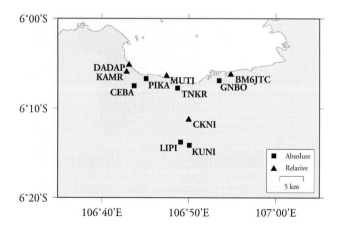

Fig. 5.8 Gravity points in Jakarta

Fig. 5.9   A10 measurements in the field

absolute gravity data in 2008. After the 2008 survey, A10-017 was returned to MGL for overhaul, and then the 2nd ion pump was installed for upgrading the vacuum capability. In the 2009 survey, the vacuum problem has been settled. However we found another problem that the laser and other controls were unstable in high temp circumstances. Fortunately these problems were withstood by cooling down the instrument and we obtained rather good absolute gravity data of about 10 µgal accuracy.

Due to the luck of the absolute gravity data in 2008, we have not yet obtained the data of enough reliable gravity changes in Jakarta. Nevertheless the result of relative gravity measurements suggested the gravity increases in the coastal area where the large subsidence was observed by GPS. We plan to conduct the same measurements in 2010 and then we expect more quantitative interpretation will be possible with GPS and groundwater level data.

5.3   Satellite Gravimetry for Hydrology

5.3.1   GRACE Mission

GRACE launched in March 2002 has realized the Low-Low Satellite to Satellite Tracking (L-L SST) configuration to measure the Earth's gravity fields for the first time in the history (Tapley et al. 2004a). GRACE consists of two Low Earth Orbiter (LEO) satellites at about 450 km altitude and their separation of about 200 km. The distance between the satellites is measured by a K-Band inter-satellite radar link with an extremely high precision of the order of µm. In addition, the effect of non-gravitational forces acting on the two satellites are measured by on-board accelerometers and removed from the relative motion of the satellites. Consequently information on the Earth's gravitational field can be obtained from the distance between the satellites. GRACE has enabled us to detect the gravity changes associated with various geophysical phenomena, for instance, terrestrial water storage variations in the major river basins in the world. In the following sections, we briefly describe the GRACE data and an overview of the data processing.

5.3.1.1 GRACE Data

The GRACE data are handled by the GRACE Science Data System (SDS) which is a shared system between the Jet Propulsion Laboratory (JPL), the University of Texas Center for Space Research (UTCSR) and the GeoForschungsZentrum Potsdam (GFZ). The level-0 data are the raw data of GRACE satellites received by the Raw Data Center (RDC) of the Mission Operation System (MOS). The SDS preprocesses the level-0 data and generates the level-1A data which consist of 10 Hz K-Band phases, accelerometer, star camera, GPS data, housekeeping data and so on. The level-0 and level-1A have not been opened to public because it is actually impossible to handle them by general users.

The SDS of JPL and GFZ generate the level-1B data from the Level-1A data so that the general users can handle them. The level-1B data include not only the K-Band ranging, satellite positions, accelerometer and other observation data but also all the geophysical corrections necessary for the later processing, i.e., the tides, atmospheric pressures, ocean loading effects and so on.

The level-1B data include all the necessary information to reveal the Earth's gravity field. However they are essentially the ranging data between the satellite along the orbits and it is not an easy task for the general users to retrieve the gravity field information from it. Therefore the three SDS data centers of JPL, UTCSR and GFZ calculate the spherical harmonic coefficients of the Earth's gravity fields (the Stokes coefficients) from the level-1B data for every 1 month and release them as the level 2 datasets. Each of the data centers employs slightly different software and different geophysical corrections. Furthermore, each data center occasionally reprocesses the datasets due to the improvements of the processing methods and/or upgrade of the software. Consequently there are several different versions of the level 2 datasets (e.g., Watkins 2007; Bettadpur 2007; Flechtner 2007). According to the data center and the revision number, the level 2 datasets are usually labeled such as UTCSL RL (Release)-2, JPL RL-2, GFZ RL-3 and so on. All these level 2 datasets and the level-1B datasets as well are downloadable from the JPL web site.

5.3.1.2 Data Processing of the Level 2 Data

The Earth's gravitational potential V is represented by the spherical harmonic series as

$$V(r,\phi,\lambda) = \frac{GM}{a} \sum_{l=0}^{\infty} \sum_{m=0}^{l} \left(\frac{a}{r}\right)^{l+1}$$
$$P_{lm}(\sin\phi)(C_{lm}\cos(m\lambda) + S_{lm}\sin(m\lambda)), \tag{5.2}$$

where G is the gravitational constant, M is the mass of the Earth, a is the Earth's equatorial radius, (r, θ, λ) are spherical coordinates, $P_{l,m} (\sin \phi)$ is fully normalized Legendre function, and $C_{l,m}$ and $S_{l,m}$ are the Stokes coefficients which are fully normalized spherical harmonic coefficients of degree l and order m. Actually the GRACE level 2 data is a time series of $(C_{l,m}(t), S_{l,m}(t))$ provided for every 1 month. Since $(C_{l,m}(t), S_{l,m}(t))$ include the static gravity fields, the temporal variation of the gravity fields can be obtained by subtracting the static gravity fields from the original coefficients. If we take the temporal averages of the Stokes coefficients as $(av\{C_{l,m}(t)\}, av\{S_{l,m}(t)\})$, the temporal variations of the gravity field can be described as

$$\Delta C_{l,m}(t) = C_{l,m}(t) - av\{C_{l,m}(t)\}$$

and

$$\Delta S_{l,m}(t) = S_{l,m}(t) - av\{S_{l,m}(t)\}. \tag{5.3}$$

Once $\Delta C_{l,m}(t)$, $\Delta S_{l,m}(t)$ have been obtained, the surface mass variations can be calculated by the following equation (Wahr et al. 1998)

$$\Delta \sigma(\theta, \phi, t) = \frac{a\rho_{ave}}{3} \sum_{l=0}^{n} \sum_{m=0}^{l} \frac{2l+1}{1+k'_l} P_{l,m}(\cos\theta)$$
$$(\Delta C_{l,m}(t)\cos(m\phi) + \Delta S_{l,m}(t)\sin(m\phi)), \tag{5.4}$$

where ρ_{ave} is the mean density of the Earth.

Spatial Filtering

It is known that the GRACE level 2 data include large errors along the satellite orbits. They are called striping errors which have more powers in higher degrees. There are several studies to suppress this kind of errors (e.g., Swenson and Wahr 2006). Among them, a low pass filtering is one of the most easy methods and widely employed. The low pass filtering is a kind of spatial average. In the frequency domain of the spherical harmonic expansion, it means to give smaller weights to the higher degree coefficients. Practically (5.4) can be modified as

$$\Delta \sigma(\theta, \phi, t) = \frac{a\rho_{ave}}{3} \sum_{l=0}^{n} W_l \sum_{m=0}^{l} P_{l,m}(\cos\theta) \frac{2l+1}{1+k'_l}$$
$$(\Delta C_{l,m}(t)\cos(m\phi) + \Delta S_{l,m}(t)\sin(m\phi)), \tag{5.5}$$

where W_l is the weight value for degree l. For the GRACE data processing, the Gaussian filter (Jekili 1981), which is represented as

$$W(\gamma) = \frac{b}{2\pi} \frac{\exp[-b(1-\cos\gamma)]}{1-\exp(-2b)}, \quad b = \frac{\ln 2}{1-\cos(d/a)}. \tag{5.6}$$

in the space domain, is often employed. The parameter "d" in (5.6) is called the correlation distance and it characterizes the strength of the filter. The weight coefficients W_l in (5.5), which are the expansion coefficients of the Legendre function, can be given by the following recursive formula (Wahr et al. 1998):

$$W_0 = \frac{1}{2\mathbf{p}}, \quad W_1 = \frac{1}{2\mathbf{p}}\left[\frac{1+e^{-2b}}{1-e^{-2b}} - \frac{1}{b}\right], \quad W_{l+1} = -\frac{2l+1}{b}W_l + W_{l-1} \tag{5.7}$$

Time Series of the Mass Variations

Using the GRACE level 2 datasets, we can calculate the mass variations on the Earth with (5.5) and (5.7). Figure 5.10 shows an example of the mass variations at every 3 months in 2007 estimated from the UTCSR RL-4 GRACE level 2 data with the Gaussian filter with the correlation distance "d" of 600 km. Figure 5.10 well represents the seasonal variation of the global land water.

On the other hand, to estimate the time variations of TWS in a certain river basin accurately, the effects outside the area concerned should be removed as precise as possible. A filter specially designed for this purpose is called the optimal regional filter. Swenson et al. (2003) have proposed the way to design the filter which minimizes both GRACE observation errors and the so-called leakage errors which are the effects from the outside of the area concerned. Using the optimal regional filter, the mass variations $\Delta\sigma_{region}$ in the area can be calculated by the following formula

$$\Delta\sigma_{region} = \frac{1}{\Omega_{region}}\frac{a\rho_{ave}}{3}\sum_{l=0}^{l\max}\sum_{m=0}^{l}\frac{(2l+1)}{(1+k'_l)}\left(W_{lm}^C\Delta C_{lm} + W_{lm}^S\Delta S_{lm}\right) \tag{5.8}$$

where Ω_{region} is the angular area, l_{max} is the maximum degree of the spherical harmonic coefficients. $W_{lm}{}^C$ and $W_{lm}{}^S$ in (5.8) are the weights of the optimal filter, and given by

$$\begin{Bmatrix} W_{lm}^C \\ W_{lm}^S \end{Bmatrix} = \left[1 + \frac{2(2l+1)B_l^2}{\sigma_0^2 W_l(1+k'_l)^2}\left(\frac{a\rho_{ave}}{3}\right)^2\right]^{-1}\begin{Bmatrix} \vartheta_{lm}^C \\ \vartheta_{lm}^S \end{Bmatrix} \tag{5.9}$$

where B_l is the degree amplitudes of the satellite measurement errors, σ_0^2 is the local signal variance, and $\vartheta_{lm}{}^C$ and $\vartheta_{lm}{}^S$ are the spherical harmonic coefficients of the regional template, i.e., 1 for the inside and 0 for the outside of the area concerned.

For the optimal design of the regional filter, B_l in (5.9) and "d" in (5.6) should be fixed in advance, and σ_0^2 is determined so as to minimize the sum of the satellite measurement errors and the leakage errors iteratively. Finally the amplitude degradation was corrected by multiplying a factor which was given as the ratio between non-filtered and filtered model data.

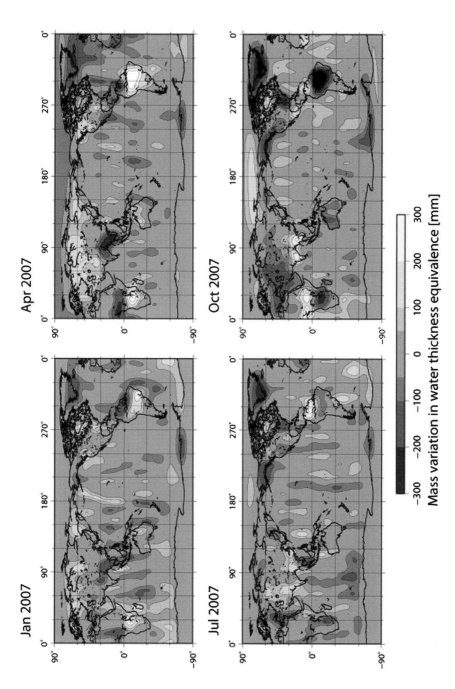

Fig. 5.10 Seasonal mass changes in 2007 observed by GRACE

5.3.2 Land Water Variation in Indochina Peninsula

After Tapley et al. (2004b) have successfully revealed the seasonal mass variations in the Amazon river basin, GRACE data have been widely employed for the studies of TWS variations in various regions in the world (e.g., Famiglietti et al. 2004; Chen et al. 2007; Yamamoto et al. 2008) . Regarding the TWS variations, the Indochina Peninsula has the 2nd largest signals in the world due to onset and offset of the East Asian monsoon. In addition, there are four major rivers, i.e., Mekong, Irrawaddy, Salween and Chao Phraya, basin sizes of which are different each other. Therefore it should be a good validation method for GRACE data to recover the mass variations in each of the basins and compare them with land water model estimations and/or other datasets.

From these points of views, Yamamoto et al. (2007) detected the mass variations in the four major river basins using the optimal regional filter of (5.8) and compared the results with a land water model of the Japan Meteorological Agency (JMA), which is a combined model of the Simple Biosphere (SiB) model (Sellers et al. 1986) and Global River flow for Total Runoff Integrating Pathways (GRiveT) model (Nohara et al. 2006). The comparisons with UTCSR RL02, JPL RL02 and GFZ RL03 data were made for each of the river basins and the combined area of the four river basins as well. Figure 5.11 shows the location of these river basins and Fig. 5.12 shows the comparison results between UTCSR RL02 and the land water model. Note that other GRACE datasets showed almost the same results. Figure 5.12 showed good agreements between the GRACE estimations and the land water model values for the Mekong, Irrawaddy basins and the combined area, while the agreements for Salween and Chao Phraya basins were poor. This was mainly due to the limitations of the spatial resolutions (about 600 km) of the GRACE data at the time.

Another important finding in Fig. 5.12 was that the phases of GRACE estimation were delayed about 1 month compared to the model variations. Although the model took into consideration snow storage, soil moisture and river storage, the groundwater storage process was only considered insufficiently. Therefore the phase differences were probably due to the improper treatments of the groundwater storage process in the hydrological model.

Considering these points, Fukuda et al. (2009) employed the JRA-JCDAS LDA and GRiveT model (JLG model) by JMA. The model included SVATS (Soil-Vegetation-Atmosphere Transfer Scheme), river flow routing, and groundwater storage models. The SVATS outputs were obtained from Japan Re-analysis 25-year data (JRA-25; Onogi et al. 2007) and the river flow routing and groundwater models were run in offline mode forced with the total runoff in the JRA-25 dataset. Assuming different treatments for the groundwater and river flow routing, several different versions of the models were estimated (Nakaegawa et al. 2007). These models were compared with an updated GRACE level 2 solution of UTCSR RL04 (Bettadpur 2007), which provides the longest data period (from April 2002 to February 2007) at that time. The comparison result showed that the best fit model was the one that include the groundwater storage.

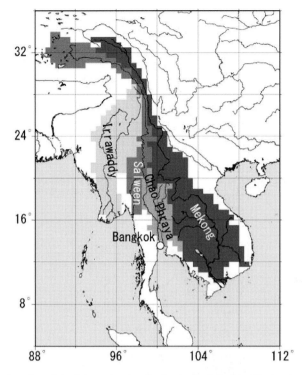

Fig. 5.11 The location of the four major river basins in the Indochina Peninsula (see Yamamoto et al. 2007)

Figure 5.13a–e depict the estimated mass variations from GRACE and the best fit model for the combined area of the four rivers (the Mekong, Irrawaddy, Salween, and Chao Phraya river basins). Large improvement in the phase differences can clearly be confirmed. This means that the GRACE data greatly contribute to modify the model parameters. In addition, the GRACE estimations in these figures exhibit basically good agreement with the model estimations, although the GRACE estimations in Salween and Chao Phraya basins are noisier than the others. There is a huge improvement in RL04 datasets compared to the previous versions (see Fig. 5.12), in which the noises in Salween and Chao Phraya basins were too large to mask almost all useful signals there.

Recently a new GRACE dataset (GRGS gravity fields RL2) has been released from the CNES (Centre National d'Etudes Spatiales)/GRGS (Groupe de Recherche en Géodésie Spatiale) group (CNES/GRGS 2009). The dataset provides every 10 days gravity field models up to degree and order 50. The models were constrained towards the static gravity field of EIGEN-GRGS.RL02.MEAN-FIELD so that the higher degree errors were suppressed. Consequently, the spatial filtering has not been necessary anymore to obtain a reliable result. Using the GRGS dataset, Yamamoto et al. (2009) detected not only seasonal mass variations but also inter-annual

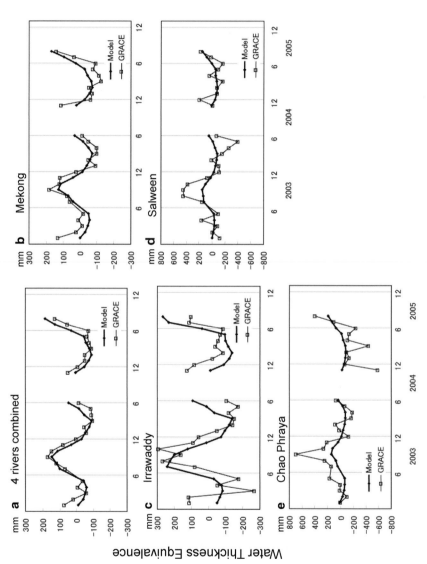

Fig. 5.12 Comparisons of the GRACE (UTCSR RL02) and the land water model (see Yamamoto et al. 2007)

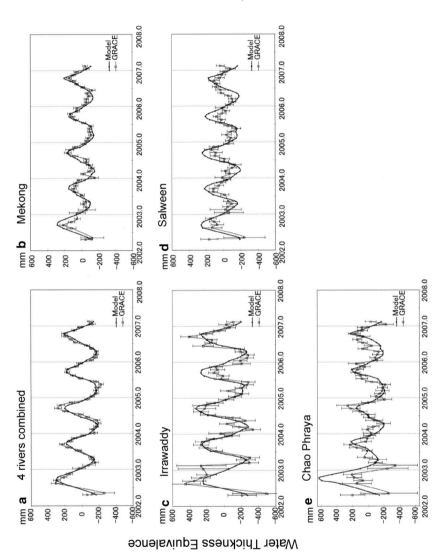

Fig. 5.13 Comparisons of the GRACE (UTCSR RL04) and the JLG land water model (see Fukuda et al. 2009)

mass trend. The GRACE estimation shows a good agreement with the land water
model down to the Chao Phraya river basin scale (178,000 km²). Moreover it shows
a clear negative mass trend over the Chao Phraya river basin, while positive trend
over the other river basins of Mekong, Irrawaddy and Salween. This tendency is
consistent with the global TWS model, but is not directly compared with an unban
scale groundwater flow model. This will be discussed again in Chap. 4.

5.4 The 2006 Australian Drought

Australia suffered historic drought attributed to rainfall deficiency in late 2006.
The Australian Government Bureau of Meteorology (BoM) reported that it was the driest
August to November period averaged across South Australia in the historical record
dating from 1900 (BoM 2006). Figure 5.14 shows the 2006 rainfall deficiency reported
by BoM. The drought made severe impacts not only on Australian human society by a
resultant reduction of the overall economic growth, but also on world food situation.

Hasegawa et al. (2008) employed the 47 monthly UTCSR RL04 datasets between
2003 and 2006 to detect the mass changes due to the drought. They first applied the
de-correlation filter (Swenson and Wahr 2006) to suppress the striping errors. In addition, they employed the Gaussian filter and the optimal regional filter for the purposes
to reveal the spatial distributions of the TWS changes and to estimate the time series
of the TWS changes, respectively. Figure 5.15 shows the mass anomaly in 2006 which
was calculated as the difference of the GRACE solutions between the average of 2006

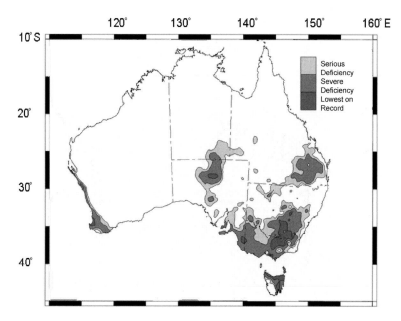

Fig. 5.14 The rainfall deficiency by the 2006 Australian Drought (from BoM 2006)

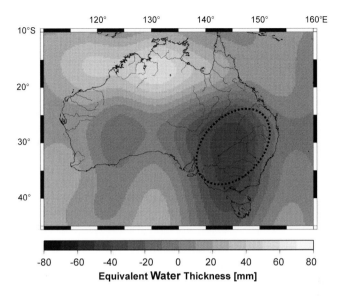

Fig. 5.15 The mass anomaly revealed by the UTCSR RL04 GRACE data. The mass anomalies were calculated as the differences between the average of 2006 and the one from 2003 to 2005. The *dotted line* shows the Murray-Daring river basin where the comparisons in Fig. 5.16 have been conducted

and the one from 2003 to 2005. Corresponding to the area of the rainfall deficiency in Fig. 5.14, the clear negative mass anomalies are found in Southeast Australia.

For the comparison between GRACE and land water models, an optimal regional filter was designed so as to extract the GRACE TWS variations in the Murray Darling River basins where the most significant negative mass anomalies were detected. The location of the Murray-Daring river basins is shown by the dotted line in Fig. 5.15.

The land water models employed are the JLG model by JMA, and the Global Land Data Assimilation System model (the GLDAS model) developed by NASA. The GLDAS model employs Mosaic land surface model with observation-based meteorological fields as atmospheric force (Rodell et al. 2004).

Figure 5.16 shows the TWS variations in the Murray-Daring river basins observed by GRACE and estimated from the land water models. Note that the annual and the semi-annual variations were removed in advance to highlight the inter-annual variations. Figure 5.16 shows that the model estimations of the TWS anomaly in 2006 are much smaller than the GRACE observation. This would suggest that the hydrological models may not properly recover the TWS changes caused by the drought. Hydrological models possibly contain various factors which may cause the errors in TWS estimations, while the GRACE data might contain some errors which cause over-estimations. It is true that there remain large uncertainties in the model estimations, for instance, in forcing fields, model structure, process description, and parameterization.

Hasegawa et al. (2009) have recently obtained the same results using the CNES/ GRGS datasets that the model estimations were too small compared to the GRACE data. They have also confirmed that the GRACE data was consistent with the in-situ

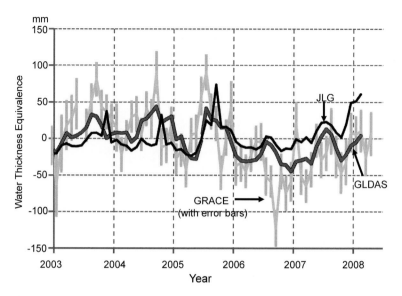

Fig. 5.16 The comparisons between the GRACE observations and the land water model estimations

superconducting gravity data in Canberra. All these demonstrated that the GRACE data, even which may contain some uncertainties, provide not only seasonal TWS changes but also the secular TWS changes which can be considered the response to the long-term climate changes. We expect that the GRACE data should make great contribution to future understanding of climate impacts to water resource and its managements.

5.5 Role of Hydrological Models

It is obvious that there are large differences in temporal and spatial resolutions between satellite and in-situ gravity measurements. In general, these differences could be resolved by hydrological models and the issue about the down-scaling (up-scaling) of the models was discussed in Fukuda et al. (2009). As described in 3.2, GRACE and the global TWS (JLG) model show a negative mass trend in Bangkok located in the Chao Phraya downstream. Bangkok is a target city of the RIHN project 2.4 and there are other regional-local hydrological models built up in different schemes and spatial scales. Yamanaka and Mikita (2009) quantitatively estimated the recharge and renewal of the confined groundwater in Bangkok using a three-dimensional groundwater flow model. Their result shows a slight positive trend in the deep groundwater storage in Bangkok, which is consistent with the observed groundwater levels of the confined groundwater wells. On the other hand, Tanaka et al. (2009) estimated the shallow aquifer recharge in Bangkok using the SWAT (Soil and Water Assessment Tool) model with detailed land use, soil map, climate conditions and topographic data. Their result shows significantly reduced recharge to the shallow aquifer, considerably larger than the GRACE estimation.

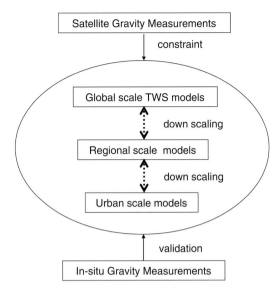

Fig. 5.17 A schematic view which shows the relations between gravity measurements and hydrological models

These inconsistencies would not be surprising because the hydrological models have not been constrained properly in terms of the total mass changes within the area concerned. Actually there was no method to measure the total mass changes before GRACE mission. In the future, the mass conservation in a local model would be constrained by a global model validated by GRACE. Moreover in-situ gravity measurements would be served as a validation tool of the local model. Conversely the models should be necessary to reveal the mechanism of the groundwater changes because gravity methods only provide the information of the mass variations. If the models nested from global to local scales can explain both satellite and in-situ gravity observations, it means that the mechanism of the groundwater movements are well understood. Figure 5.17 illustrates such relations between models and observations. It would be an ideal situation that the gravity methods can contribute to monitor the groundwater variations from the urban to global spatial scales.

5.6   Concluding Remark

The sustainable use of groundwater is a key issue for future urban developments, and the gravity techniques, which are new and still challenging, should contribute to monitor the groundwater variations, because, as described repeatedly, only gravity measurements can detect the mass variations directly. This is the most important point that we have employed gravity measurements on land and from space as well for the studies. The gravity data can provide the most basic information to manage the urban water usability together with other hydrological information such as the groundwater levels measured at observation wells.

Regarding the ground base gravity measurements, relative gravimeters have been usually employed so far. However, relative measurements always include some uncertainties in interpretation processes, and we consider that the use of absolute gravimeters is a key for these studies. As previously described, the A10 gravimeter enables the absolute gravity measurements in the field with much ease. We expect such measurements become more popular in the future.

The spatial resolution of the GRACE data is far from satisfactory for discussing urban scale groundwater variations. However, GRACE provides very good constraint for hydrological models in terms of total mass variations. It should be noted again that there was no such high precision constraints as the mass variations before GRACE. In the future, several different scale models would combine much advanced in-situ and satellite observations for a better understanding of the hydrological processes.

References

Abe M, Takemoto S, Fukuda Y, Higashi T, Imanishi Y, Iwano S, Ogasawara S, Kobayashi Y, Takiguchi H, Dwipa S, Kusuma DS (2006) Hydrological effects on the superconducting gravimeter observation in Bandung. J Geodyn 41:288–295. doi:10.1016/j.jog.2005.08.030

Abidin HZ, Andreas H, Djaja R, Darmawan D, Gamal M (2008) Land subsidence characteristics of Jakarta between 1997 and 2005, as estimated using GPS surveys. GPS Solut 12:23–32. doi:10.1007/s10291-007-0061-0

Allis RG, Hunt TM (1986) Analysis of exploitation induced gravity changes at Wairakei geothermal field. Geophysics 51:1647–1660

Bettadpur S (2007) UTCSR level-2 processing standards document for level-2 product release 0004, GRACE 327-742 (CSR-GR-03-03). Center for Space Research, The University of Texas at Austin, Austin

BoM (Bureau of Meteorology) (2006) Drought statement issued 4th December. http://www.bom.gov.au/announcements/media_releases/climate/drought/20061204.shtml

Bower D, Courtier N (1998) Precipitation effects on gravity measurements at the Canadian absolute gravity site. Phys Earth Planet Int 106:353–369. doi:10.1016/S0031-9201(97) 00101-5

Chen JL, Rodell M, Wilson CR, Famiglietti JS (2005) Low degree spherical harmonic influences on Gravity Recovery and Climate Experiment (GRACE) water storage estimates. Geophys Res Lett 32:L14405. doi:10.1029/2005GL022964

Chen JL, Wilson CR, Famiglietti JS, Rodell M (2007) Attenuation effect on seasonal basin-scale water storage changes from GRACE time-variable gravity. J Geodyn 81:237–245. doi:10.1007/s00190-006-0104-2

CNES/GRGS (2009) Time-variable gravity fields (RL02). http://bgi.cnes.fr:8110/geoid-variations/RL02.html

Famiglietti J, Chen J, Rodell M, Seo K, Syed T, Wilson C (2004) Terrestrial water storage variations from GRACE. Presented at IAG international symposium, gravity, geoid and space missions – GGSM2004, 2004, Porto

Flechtner F (2007) GFZ level-2 processing standards document for level-2 product release 0004, GRACE 327-743 (GR-GFZ-STD-001). GeoForschungszentrum Potsdam, Wessling

Fukuda Y, Higashi T, Takemoto S, Abe M, Dwipa S, Kusuma DS, Andan A, Doi K, Imanishi Y, Arduino G (2004) The first absolute gravity measurements in Indonesia. J Geodyn 38:489–501

Fukuda Y, Yamamoto K, Hasegawa T, Nakaegawa T, Nishijima J, Taniguchi M (2009) Monitoring groundwater variation by satellite and implications for in-situ gravity measurements. Sci Total Environ 407:3173–3180. doi:10.1016/j.scitotenv.2008.05.018

Hasegawa T, Fukuda Y, Yamamoto K, Nakaegawa T (2008) The 2006 Australian drought detected by GRACE. In: Taniguchi M et al (eds) Headwaters to the ocean. Taylor & Francis Group, London, pp 363–367, ISBN 978-0-415-47279-1

Hasegawa T, Fukuda Y, Yamamoto K, Nakaegawa T, Tamura Y (2009) Long-term trends of terrestrial water storage in south-east Australia detected by GRACE. In: The 3rd international symposium, human impacts on urban subsurface environment, 17–20 November 2009, Abstract, p 22

Jekili C (1981) Alternative methods to smooth the Earth's gravity field: Rep. 327. Department of Geodetic Science and Survey, Ohio State University, Columbus

Lambert A, Beaumont C (1977) Nano variations in gravity due to seasonal groundwater movements: Implications for the gravitational detection of tectonic movements. J Geophys Res 82:297–306

MGL (Micro-g LaCoste Inc.) (2008) A-10 portable gravimeter user's manual. http://www.microglacoste.com/pdf/A-10Manual.pdf, 58 pp

Nakaegawa T, Tokuhiro T, Itoh A, Hosaka M (2007) Evaluation of seasonal cycles of hydrological processes in Japan meteorological agency land data analysis. Pap Meteorol Geophys 58:73–83. doi:10.2467/mripapers.58.73

Nishijima J, Fujimitsu Y, Ehara S (2007) Geothermal reservoir monitoring using repeat gravity measurement at Takigami geothermal field, central Kyushu, Japan. In: Proceedings of Renewable Energy 2006 (CD-ROM).

Nohara D, Kitoh A, Hosaka M, Oki T (2006) Impact of climate change on river discharge projected by multimodel ensemble. J Hydromet 7:1076–1089. doi:10.1175/JHM531.1

Nordquist G, Protacio JAP, Acuna JA (2004) Precision gravity monitoring of the Bulalo geothermal field, Philippines: independent checks and constraints on numerical simulation. Geothermics 33:37–56

Onogi K, Tsutsui J, Koide H, Sakamoto M, Kobayashi S, Hatsushika H, Matsumoto T, Yamazaki N, Kamahori H, Takahashi K, Kadokura S, Wada K, Kato K, Oyama R, Ose T, Mannoji N, Taira R (2007) The JRA-25 reanalysis. J Meteor Soc Japan 85:369–432

Rodell M, Houser PR, Jambor U, Gottschalck J, Mitchell K, Meng CJ, Arsenault K, Cosgrove B, Radakovich J, Bosilovich M, Entin JK, Walker JP, Lohmann D, Toll D (2004) The global land data assimilation system. Bull Am Meteor Soc 85(3):381–394

Schemerge D, Francis O (2006) Set standard deviation, repeatability and offset of absolute gravimeter A10-008. Metrologia 43:414–418. doi:10.1088/0026-1394/43/5/012

Schepers JHG (1926) De Primaire Kringen Ia En II, Benevens Het. Stadsnet Van Batavia En Weltevreden, No.1 De Nauwkeurrig-heidswaterpassing Van Java. Topografische Dienst In Neder-lansch-Indie. Weltevreden Reproductiebedrifj Top. Dienst t, 64 pp, with loose collection of maps in pocket at end.

Sellers PJ, Mintz Y, Sud YC, Dalcher A (1986) Simple Biosphere model (SiB) for use within general circulation model. J Atmos Sci 43:505–531

Sofyan Y (2009) Repeat gravity measurement for groundwater level monitoring – an application to the reservoir monitoring in geothermal power plant. In: The 3rd international symposium, human impacts on urban subsurface environment, 17–20 November 2009, Abstract, p 23

Suharto P (1971) Precise leveling in Indonesia. Final undergraduate report. Department of Geodetic Engineering, Institute of Technology, Bandung

Swenson S, Wahr J (2006) Post-processing removal of correlated errors in GRACE data. Geophys Res Lett 33:L08402. doi:10.1029/ 2005GL025285

Swenson S, Wahr J, Milly PCD (2003) Estimated accuracies of regional water storage variations inferred from the Gravity Recovery and Climate Experiment (GRACE). Water Resour Res 39:1223. doi:10.1029/2002WR001808

Tanaka et al (2009) The effects of urbanization on shallow aquifer recharge in Asian megacities: an application of the SWAT model to Bangkok. In: The 3rd international symposium, human impacts on urban subsurface environment, 17–20 November 2009, Abstract, p 4

Taniguchi M, Shimada J, Fukuda Y, Onodera S, Yamamo M, Yoshikoshi A, Kaneko S (2008) Degradation of subsurface environment in Asian coastal cities. In: Taniguchi M et al (eds) Headwaters to the ocean. Taylor & Francis Group, London, pp 605–610, ISBN 978-0-415-47279-1

Tapley BD, Bettadpur S, Watkins M, Reigber C (2004a) The Gravity Recovery and Climate Experiment: mission overview and early results. Geophys Res Lett 31:L09607. doi:10.1029/2004GL019920

Tapley BD, Bettadpur S, Ries JC, Thompson PF, Watkins MM (2004b) GRACE measurements of mass variability in the earth system. Science 203:503–505

Wahr J, Molenaar M, Bryan F (1998) Time-variability of the Earth's gravity field: hydrological and oceanic effects and their possible detection using GRACE. J Geophys Res 103(B12):30205–30230

Watkins MM (2007) JPL level-2 processing standards document for level-2 product release 04, GRACE 327-744. Jet Propulsion Laboratory, Pasadena

Yamamoto K, Fukuda Y, Nakaegawa T, Nishijima J (2007) Landwater variation in four major river basins of the Indochina peninsula as revealed by GRACE. Earth Planet Space 59:193–200

Yamamoto K, Hasegawa T, Fukuda Y, Nakaegawa T, Taniguchi M (2008) Improvement of JLG terrestrial water storage model using GRACE satellite gravity data. In: Taniguchi M et al (eds) Headwaters to the ocean. Taylar & Francis Group, London, pp 369–374, ISBN 978-0-415-47279-1

Yamamoto K, Fukuda Y, Nakaegawa T, Hasegawa T, Taniguchi M (2009) Study on landwater variation over Chao Phraya river basin using GRACE satellite gravity data. In: The 3rd international symposium, human impacts on urban subsurface environment, 17–20 November 2009, Abstract, p 21

Yamanaka T, Mikita M (2009) Disturbance of groundwater flow system due to excessive pumping in the Bangkok metropolitan area, Thailand. In: The 3rd international symposium, human impacts on urban subsurface environment, 17–20 November 2009, Abstract, p 14

Chapter 6
The Proposed Groundwater Management for the Greater Jakarta Area, Indonesia

Robert M. Delinom

Abstract The Greater Jakarta, the capital city of Republic of Indonesia, occupies the northern zone of Java Island and the elevations of this plain vary from 0 to 1,000 m above sea level. Some evident showed that the population influenced the condition of groundwater resources in this megacity as most of them acquire the needed water by groundwater abstraction. It caused a negative impact on these resources itself both quantity and quality. Therefore, the proper groundwater management of this area should be established and the groundwater management should cover the two important aspects, i.e., physical and technical aspects, and social and non technical aspects.

6.1 Introduction

Since the beginning of the twentieth century, groundwater of the Greater Jakarta basin has been used for drinking water and other water resources purposes. Unfortunately, groundwater use is increasing year by year and some problems are threatening this fragile aquifer system. It has influenced either quality or quantity of groundwater (Delinom et al. 2009).

The dependency of industry on groundwater is one of the constraints that faced by groundwater management effort. It is associated with the lack of infrastructure that provided by the government. According to the most recent data, the amount of clean surface water that supplied to the industrial sector was only about 3.5 million m^3 in 2003, which is just 1% of the volume required by industry (Statistical Local Office of Greater Jakarta 2003). This means that almost all water required by the industrial sector comes from groundwater. The trend of groundwater tendency due to groundwater abstraction in Jakarta Area is shown on Fig. 6.1.

R.M. Delinom (✉)
Research Center for Geotechnology, Indonesian Institute of Sciences, Cisitu Campus,
Jln. Cisitu – Sangkuriang, Bandung 40135, Indonesia
e-mail: rm.delinom@geotek.lipi.go.id; delinom2002@yahoo.com

M. Taniguchi (ed.), *Groundwater and Subsurface Environments: Human Impacts in Asian Coastal Cities*, DOI 10.1007/978-4-431-53904-9_6, © Springer 2011

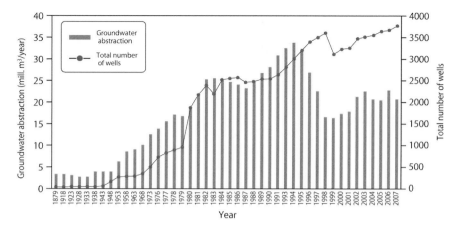

Fig. 6.1 Trend of groundwater abstraction in Jakarta Area (Statistical Local Office of Greater Jakarta 2003)

Another factor influencing the scarcity of groundwater is the condition of groundwater recharge area. Groundwater recharge can be interpreted as the addition to the groundwater from an external area to the saturated water column. Generally, groundwater is replenished from rainfall, rivers and human intervention such as an artificial recharge well or lake. One of the main factors influencing groundwater depletion is significant changes of the land cover from natural terrain to the developed areas, especially in the recharge area.

6.2 The Study Area

The Greater Jakarta is the capital city of Republic of Indonesia. It occupies the northern zone of Java Island that comprises low hilly areas of folded Tertiary strata, and Quaternary coastal lowlands bordering the Java Sea (Fig. 6.2). Two Quaternary formations and three young Tertiary formations act as groundwater aquifers zone and one quaternary formation act as an aquitard. Some older formations present as basement of the basin. The elevations of this plain vary from 0 to 1,000 m above sea level. It is one of the most developed basins in Indonesia and is located between 106° $33'$–107° E longitude and 5° $48'$ $30''$–6° $10'$ $30''$ S latitude covering an area of about 652 km². It has a humid tropical climate with annual rainfall varying between 1,500 and 2,500 mm and is influenced by the Monsoon.

The population of Jakarta at present is around 7.5 millions (Jakarta Local Government Website 2007). It represents the official number of population actually living in the Greater Jakarta Area. The reality which is faced by Jakarta is that many people who are working in Jakarta during the daytime are living in the adjacent cities i.e., Bogor, Depok, Tanggerang, and Bekasi (Bodetabek Area). Since the operation of the Jakarta – Bandung Highway, some people living in the cities of

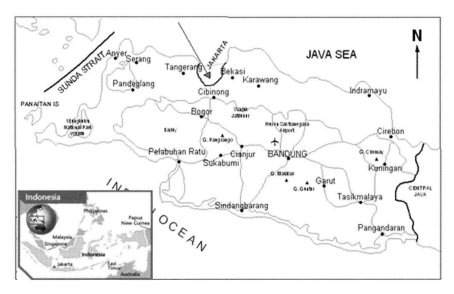

Fig. 6.2 Location Map of the Greater Jakarta. It is the Capital City of Republic of Indonesia and located in the coastal area of Java Island (Copied from Delinom 2008)

Purwakarta and Bandung have also become commuters. This circumstance has caused the population of Jakarta to increase up to 10 or 11 millions during the weekdays. It is obvious that urbanization has increased the water demand in this area. As surface water provides covers only 30% of water demand, people are harvesting the available groundwater in the basin. In Jakarta Groundwater Basin, the use of groundwater has greatly accelerated conforming to the rise in its population and the development of industrial sector, which consume a relatively huge amount of water.

According to Engelen and Kloosterman (1996), structurally, the Jakarta groundwater basin is part of the so called a Northern Zone comprising the low hilly areas of folded Tertiary strata, and coastal lowlands bordering the Java Sea.

Geologically, the study area is dominated by Quaternary sediment and, unconformably, the base of the aquifer system is formed by impermeable Miocene sediments which are cropping out at the southern boundary, which were known as Tanggerang High in the west, Depok High in the middle and Rengasdengklok High in the east. They acted as the southern basin boundary. The basin fill, which consist of marine Pliocene and Quaternary sand and delta sediments, is up to 300 m thick. Individual sand horizons are typically 1–5 m thick and comprise only 20% of the total fill deposits. Silts and clays separate these horizons. Fine sand and silt are very frequent components of these aquifers (Martodjojo 1984; Assegaf 1998).

In detail, Sudjatmiko (1972), Effendi et al. (1974) and Turkandi et al. (1992) differentiated the lithology that cropping out in this area into some lithologies and explained as follows (Fig. 6.3):

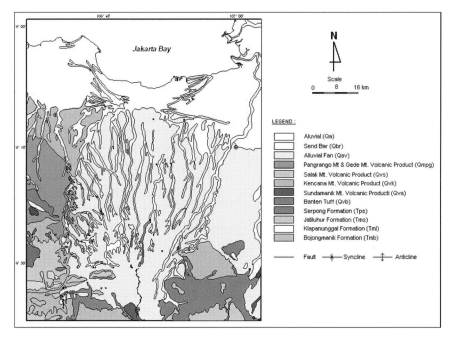

Fig. 6.3 Geological Map of the Greater Jakarta and its surrounding area. At the surface, the litho logy were dominated by coastal and deltaic deposits (Effendi et al. 1974; Sudjatmiko 1972; Turkandi et al. 1992). Copied from Delinom et al. 2009

1. Banten Tuff, developed by young volcanic eruptive material.
2. Bogor Fan, consists of fine tuff, sandy tuff intercalated with conglomerate as result of Mount Gede – Pangrango volcanic activity.
3. Paleo and recent beach ridge deposits which are deposited parallel to recent coastal line.
4. Alluvium deposits consist of silt, sand, and gravel. In some parts, this deposit was covered by river sediment that composed by gravel, sand, silt and clay. The remnant of vegetation was found at a certain depth.

Based on boreholes analysis, the formation were found in subsurface are:

1. *Rengganis Formation* consists of fine sandstones and clay stone outcropped in the area of Parungpanjang, Bogor. Un-conformably, this formation is covered by coral limestone, marl, and quartz sandstone.
2. *Bojongmanik Formation*, consist of interbedded of sandstone and clay stone, with intercalated limestone.
3. *Genteng Formation*, consist of volcanic eruption material such as andesitic breccias and intercalated tuffaceous limestone.
4. *Serpong Formation*, interbedded of conglomerate, sandstone, marl, pumice conglomerate, and tuffaceous pumice.

6.3   Methodology

To formulate a proper groundwater management in an urban area, the assessment of land zoning and groundwater hazard of the area must be identified. Land zoning assessment will illustrate closely the area of groundwater sources (recharge area), groundwater aquifer, groundwater use (discharge area), and groundwater vulnerability (environmental sensitivity). The identification of groundwater hazard assessment dealt with groundwater quality and groundwater quantity which are hook up with physical and social aspects. The hazard potential in recharge and discharge area should be recognized before hazard management is determined. The combination of both aspects, land zoning and groundwater hazard, will be the proper groundwater management of the area. Scheme of groundwater management is presented on Fig. 6.4.

The main threats to groundwater sustainability arise from the steady increase in water demand and from the increasing use and disposal of chemicals to the land surface. Management is required to avoid serious degradation and there needs to be increased awareness of groundwater at the planning stage, to ensure equity for all stakeholders and most important of all to match water quality to end use. Despite the threats from potentially polluting activities, groundwater is often surprisingly resilient, and water quality over large area of the world remains good. A vital aid to good groundwater management is a well-conceived and properly supported monitoring and surveillance system. For this reason monitoring systems should be periodically reassessed to make sure that they remain capable of informing management decisions so as to afford early warning of degradation and provide valuable time to devise an effective strategy for sustainable management.

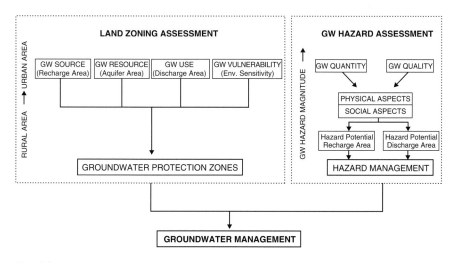

Fig. 6.4 Scheme of groundwater management in Greater Jakarta Area (Delinom 2008)

To improve the groundwater management, the sustainable groundwater management strategy should be employed. This strategy covers long term groundwater resources conservation, groundwater quality protection; change the groundwater resources management paradigm to groundwater as a non renewable resource.

The future challenge for groundwater management is to alter the mechanism of water provision that currently applies. In addition the changing environment as consequence of the development has also brought undesirable effects to the quantity of groundwater. In order to manage the groundwater potential in its optimal capacity, it is important to identify exactly where the recharge area take place and which quantities are involved.

It was recognised that the groundwater problems in recharge area is different with the groundwater problems in discharge area. The main groundwater problem in recharge area is the decreased of groundwater recharge which is caused by land use degradation. This substance can initiate the runoff increased and groundwater storage decreased, and creates flood and drought disasters. Therefore, the recharge area management should be employed appropriately. The main problem in discharge area is the increased groundwater usage for human activities. It causes groundwater table descent and groundwater storage reduction and creates land subsidence, groundwater pollution, and drought disasters. Those problems then lead to flood disaster and groundwater resources crisis. In the discharge area, the things that should be executed are groundwater abstraction management. It is known that for doing the groundwater management, the basin geometry and the cover of recharge and discharge area should be defined first.

6.4 Megacity Groundwater Properties

There are 5 (five) main factors that influences the groundwater resources in a mega city such as Jakarta i.e., global climate change, population pressure, urbanization, agricultural and industrial activities. It is known that global climate change phenomena have increased the sea water level. It influenced the position of shorelines in some parts of the world, including northern part of Jakarta area that has border with the Java Sea. Like many other cities that located on coastal area, sea water encroached into the land and influenced either surface or groundwater resources. Total of population, urbanization and industrial activities created a pressure to the groundwater resources due to groundwater over-abstraction activity to fulfil their daily needs. The urbanization can also increase the impervious cover, drains, utility lines, backfilled areas, surface flow, point sources for recharge and contamination.

The potential impacts of urbanization on groundwater resources are the resources availability and quality degradation. Some impacts of groundwater use on urbanization are infrastructure damage that is caused by the occurrence of land subsidence and infrastructure drainage and uplifted problems. Agriculture has a reciprocal relationship with the groundwater resources as it needs some groundwater resources for growing plants and in the other side, plants can act as an instrument in helping water to recharge into the soil. Public health condition is very much depending on the groundwater condition as people in Jakarta Area fulfil their water daily need from groundwater. The worse groundwater quality condition the worse public health of the area.

The groundwater in urban area is abstracted from aquifers through dug or drilling wells. Together with surface water, they are used to supply domestic, industrial, and agricultural activities. The waste water from those activities then are treated and used for irrigation or injected back to the aquifers. The urban groundwater quantity is depend on the aquifers recharged, impermeable covers, artificial replenishment to increase aquifers recharge. Some alternatives to increase the water resources are: increase in surface storage; improve of groundwater management; increase in water utilization efficiency; and large-scale inter basin water transfers.

6.5   Jakarta Groundwater Condition

Overexploitation of groundwater has become a common issue along the coastal area where good quality groundwater is available. Consequently, many coastal regions in the world experience extensive saltwater intrusion. It is obvious that urbanization has increased the water demand in this area. As the drinking water which is supplied by surface water only covers 30% of water demand, people are harvesting the available groundwater in the basin. In the Jakarta Groundwater Basin, the use of groundwater has greatly accelerated conforming to the rise of its population and the development of the industrial sector. Over-pumpage can also decrease the volume of groundwater and result in land surface subsidence. The subsurface layer compaction also supports the existence of land subsidence. Geyh and Soefner (1996) reported on the salt water intrusion phenomena in the Jakarta Area. Djaja et al. (2004) recognized land subsidence phenomenon occurring in some parts of the Jakarta Metropolitan Area.

Based on groundwater monitored data of 51 monitoring wells around Jakarta area, it can be concluded that most of water level in Jakarta area of five clusters aquifers i.e. 0–40, 40–95, 95–140, 140–190, and 190–250 m, were decreased (Delinom et al. 2009) (Fig. 6.5). The water quality analysis from 27 shallow wells that were distributed over Jakarta Area showed that the nitrate content in 2009 were very high in some places. It means that the human daily activities had influenced the water quality in this area (Sudaryanto and Suherman 2008) (Fig. 6.6).

The estimated subsidence rates during the period Dec. 1997–Sept. 2005 are 1–10 cm/year and reach 15–20 cm/year. The highest rates of land subsidence occur in northwestern Jakarta. The central and north-eastern parts occasionally also show quite high rates of subsidence. These vertical temporal variations however, may still be contaminated by annual/semiannual signal bias that plagues all GPS temporal measurements (Abidin et al. 2007). From the observation period 1982–1991, the highest subsidence occurred at Cengkareng (North Jakarta) with a rate of 8.5 cm/year. In the period 1997–1999, the highest subsidence occurred at Daan Mogot (North-west Jakarta) with a rate of 31.9 cm/year. The rate increase shows that the land subsidence in Jakarta is continuing.

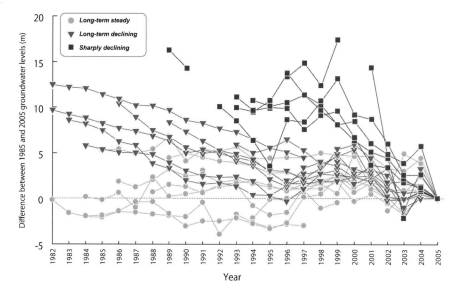

Fig. 6.5 Groundwater level fluctuations between of 2001 and 2005 at some locations in Greater Jakarta Area. It is showed that the groundwater tend to decrease

6.6 The Proposed Groundwater Management in Greater Jakarta Area

The groundwater management problem in the Jakarta Basin has many dimensions, one of them is to provide alternative source of water for industrial use. Looking at the groundwater control mechanism in the Jakarta Basin, licensing is still considered the main tool for controlling groundwater abstraction. This mechanism would not work with the bare minimum awareness of the stakeholders about the importance of groundwater conservation and weak law enforcement and monitoring. The fact is that in the Jakarta Basin, many unregistered deep wells still have been found. There are no incentives such as tax compensation for industries that used recycle water. The result is that many industries are not interested in water conservation, making it extremely difficult to control groundwater extraction in the Jakarta Basin. The failure of water utilities to supply raw water and to extend the coverage area has also become a trigger for the groundwater problems. Industry still depends on groundwater, and since industries are self-regulating, groundwater control becomes difficult.

Based on the groundwater hazard assessment in Greater Jakarta Area, as the discharge area, the quality hazards that were found are mostly the water pollution of domestic waste and industrial activities. Quantitatively, when groundwater level and reserve decreased, the land-subsidence, flooding and drought disasters, and sea water intrusion were discovered. In the recharge area, southern part of Jakarta, the domestic waste and agricultural activities influenced the groundwater quality condition. The decline of water recharged and groundwater reserve, the increasing run

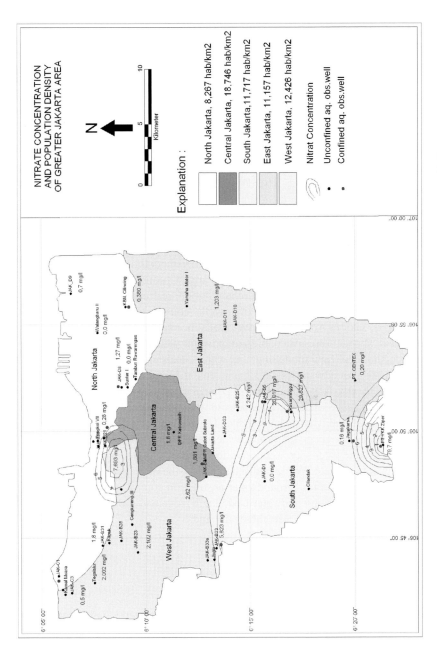

Fig 6.6 Relationship between population density and nitrate content in Jakarta Area (Sudaryanto and Suherman 2008)

Table 6.1 Groundwater risk assessment in Jakarta groundwater basin

Location	Groundwater hazard assessment	
	Groundwater quality	Groundwater quantity
Jakarta groundwater basin	– Industrial activity pollution – Domestic waste pollution – Liquid waste infiltration – Seawater intrusion – Paleo-salt leaching	– Groundwater rebound flow – Groundwater level regression (land subsidence) – Groundwater flow change – Surface runoff increment (flooding and draught disaster)
	Groundwater hazard magnitude	

off and reduced springs were encountered in this area. It is recognized that qualitatively the groundwater in Jakarta Area had been disturbed since it was infiltrated in the recharge area. Therefore, the condition of recharge area must be managed simultaneously with the discharge area.

The groundwater management should cover the two important aspects, i.e., physical and technical aspects, and social and non technical aspects. Physically and technically, for managing groundwater quality, water treatment, waste management, wells monitoring, and groundwater quality modelling should be executed both in Greater Jakarta Area and in recharge area. Groundwater quantity management in recharge area will cover land rehabilitation, re-forestation, springs conservation, artificial recharge and injection wells construction, and recharge area broadening. In Jakarta Area, it will cover wells monitored, groundwater maximum depletion and abstraction, sustainable groundwater yield determination, groundwater balance and local flow modelling, and water canals construction.

Socially, the groundwater quality management in recharge area and discharge area, Greater Jakarta Area, should cover control of groundwater source conservation zone; socialization of dangerous and environmental friendly substances utilization, groundwater quality basic knowledge. The groundwater quantity management in recharge area will cover the groundwater basic knowledge socialization, built area control, groundwater source conservation zone control, recharge area plan control, and law enforcement. While in discharge area, the discharge area plan control, groundwater abstraction tax, groundwater abstraction control, groundwater condition change monitoring, groundwater basic knowledge and sanitary system socialization, and law enforcement, see Tables 6.1 and 6.2.

6.7 Concluding Remarks

Some remarks concerning the water management in Greater Jakarta area can be indicated, among others are:

1. To improve the groundwater management, the sustainable groundwater management strategy should be employed. This strategy covers long term groundwater resources conservation, groundwater quality protection; change the

Table 6.2 Proposed groundwater management in Jakarta Groundwater Basin

Location	Proposed groundwater management			
	Physical/technical aspects		Social/nontechnical aspects	
	Groundwater quality	Groundwater quantity	Groundwater quality	Groundwater quantity
Jakarta groundwater basin	– Waste management – Water treatment – Infiltration well/injection well – Monitored wells – Groundwater quality modeling – Contaminant loading capacity – Groundwater vulnerability analysis	– Land rehabilitation – Reforestation – Springs conservation – Artificial recharge and injection well construction – Monitored wells water level – Determination of maximum groundwater depletion – Local gw-recharge analysis – Determination of sustainable yield – Water canals construction	– Regulation of groundwater resources conservation – Socialization of using dangerous substance – Socialization of using environmental friendly substance – Socialization of groundwater quality basic knowledge	– Regulation of city space – Law enforcement – Groundwater abstraction tax – Groundwater abstraction control – Groundwater condition change monitoring – Groundwater basic knowledge – Good sanitary system socialization

groundwater resources management paradigm to groundwater as a non renewable resource.

2. Greater Jakarta Area is occupied by discharge area, while the recharge area located just in the southern part of this area. Facing this reality, the groundwater management in this area must be more concerned to the problems that are discovered within discharge area.

3. Based on the groundwater hazard assessment in Greater Jakarta Area, as the discharge area, the quality hazards that were found are mostly the water pollution of domestic waste and industrial activities.

4. It is recognized that qualitatively the groundwater in Jakarta area had been disturbed since it was infiltrated the recharge area. Therefore, the condition of recharge area must be managed simultaneously with the discharge area.

5. The groundwater management should cover the two important aspects, i.e., physical and technical aspects, and social and non technical aspects.

Acknowledgement The author wishes to thank The Project of Human Impact to Subsurface Environment in Some Megacities in Asia of Research Institute for Humanity and Nature (RIHN), Kyoto, c.q. Prof. Dr. Makoto Taniguchi and Competitive Program Project of Indonesian Institute of Sciences (LIPI), c.q. Prof. Dr. Lukman Hakim, for supporting this research. He also would like to thank Makoto Kagabu, Sudaryanto, and Rachmat Fajar Lubis for their helps during the preparation of this manuscript.

References

Abidin HZ, Andreas H, Djaja R, Darmawa D, Gamal M (2007) Land subsidence characteristics of Jakarta between 1997 and 2005, as estimated using GPS surveys. GPS solutions. Springer, Berlin/Heidelberg. doi:10.1007/s10291-007-0061-0. Website: http://dx.doi.org/10.1007/s10291-007-0061-0

Asseggaf A (1998) Hidrodinamika Airtanah Alamiah Cekungan Jakarta. Msc Thesis, Dep. Teknik Geologi ITB, Bandung, Indonesia

Delinom RM (2008) Groundwater management issues in the Greater Jakarta area, Indonesia. In: Proceedings of international workshop on integrated watershed management for sustainable water use in a humid tropical region, JSPS-DGHE joint research project, vol 8, pp 40–54

Delinom RM, Assegaf A, Abidin HZ, Taniguchi M, Suherman D, Lubis RF, Yulianto E (2009) The contribution of human activities to subsurface environment degradation in Greater Jakarta Area, Indonesia. Sci Total Environ 407:3129–3141

Djaja R, Rais J, Abidin HZ, Wedyanto K (2004) Land subsidence of jakarta metropolitan area. In: Proceeding 3rd IG regional conference, Jakarta, Indonesia

Effendi AC, Kusnama K, Hermanto B (1974) Geological map of Bogor Area, Jawa, Indonesia. Geological Research and Development Center, Bandung, Indonesia

Engelen GB, Kloosterman FH (1996) Hydrological system analysis, method, and application. Kluwer, Dorddrecht, pp 140–144

Geyh MA, Soefner B (1996) Groundwater mining study by simplified sample collection in the Jakarta Basin aquifer, Indonesia. International Atomic Energy Agency. Proceedings series, pp 174–176

Jakarta Local Government Website (2007) http://dkijakarta.go.id/

Martodjojo S (1984) Evolusi Cekungan Jawa Barat. Doctor dissertation, Bandung Institute of Technology, Bandung, Indonesia

Sudaryanto M, Suherman D (2008) Degradasi Kualitas Airtanah Berdasarkan Kandungan Nitrat di Cekungan Jakarta. Jurnal RISET Geologi dan Pertambangan 18(2)

Sudjatmiko S (1972) Peta Geologi Lembar Cianjur, Jawa. Geological Research and Development Center, Bandung

Statistical Local Office of Greater Jakarta (2003) Jakarta Dalam Angka tahun 2003. BPS, Jakarta

Turkandi T, Sidarto S, Agustyanto DA, dan Hadiwidjoyo MMP (1992) Peta Geologi Lembar Jakarta dan Kepuluan Seribu, Jawa. Geological Research and Development Center, Bandung

Chapter 7
Review of Groundwater Management and Land Subsidence in Bangkok, Thailand

Oranuj Lorphensri, Anirut Ladawadee, and Surapol Dhammasarn

Abstract The history of groundwater use in Bangkok started in 1907. Since then, groundwater is the important source of public water supply concurrent with surface water. Both public and private sectors had freely developed groundwater for several decades before consequent affect revealed. Due to the past uncontrolled over pumping of groundwater, certain aquifers and overlying clay layer are under substantial stress, leading to serious land subsidence which at its most severe amounts to 10 cm/year (1978–1981). In certain places, with combined surface loading, this has amounted to a maximum recorded settlement of 100 cm over a 21-year period (1978–1999) and groundwater level has declined to 55 m from ground surface. The increasing of groundwater abstraction reached it maximum at 2.2 million cubic meters per day (Mm^3/d) in 1999. The mitigation actions have been required if they are to be reinstated and stabilized. The subsequent strict mitigations such as declaration of "Groundwater Critical Zone" covered large area totally seven provinces including Bangkok. The mitigation included reducing the permissible usage of registered wells, promoting public awareness in groundwater conservation, and finally implementing "No permission for groundwater development in public water supply service area" in the Bangkok metropolis. In addition, Groundwater Tariff and Groundwater Conservation Tax have been implemented. All these mitigations have been determined to control the total abstraction to meet permissible yield which has been studied at 1.25 Mm^3/d. The strict mitigations finally return good result.

7.1 Introduction

Bangkok is the capital city of Thailand and the biggest city in Southeast Asia. It is situated on the flood plain and delta of the Chao Phraya River which traverses the Lower Central Plain of Thailand. The total area including surrounding six provinces

O. Lorphensri (✉), A. Ladawadee, and S. Dhammasarn
Department of Groundwater Resources, Ministry of Natural Resources
and Environment, Bangkok, Thailand
e-mail: oranujl@hotmail.com

is approximately 10,200 km^2. Bangkok has a registered population of approximately 6.4 million residents while the greater Bangkok area has a registered population of 11 million residents (January 2008). It extended about 200 km from north to south and about 175 km from east to west. It is bounded on both east and west by mountain ranges and on the south by Gulf of Thailand. To the north, the plain is bordered by a series of small hill dividing it from the Upper Central Plain. The total area of the basin is about 20,000 km^2 with an average annual rainfall of 1,190 mm. The groundwater recharge to the basin was estimated as 3.2% of rainfall (AIT 1982). The principal sources of recharge are outcrops at north, east and west several kilometers of Bangkok.

7.2 Hydrogeology

The Bangkok metropolis is one of the fastest developing capitals of the world. It situated on the low-lying plain on very thick unconsolidated sediments. Since the rapid increase in the demand for groundwater, groundwater was substantially extracted from the deep aquifer beneath the city. It has been realized since 1970 that land subsidence could be experienced in this area resulting from the depletion of the beneath aquifer were to cause consolidation of the compressible clay deposits of unconsolidated sediments (AIT 1972).

Bangkok metropolis is entirely underlain by recent marine clay, 15–30 m in thickness, known as the Bangkok Clay (Fig. 7.1). This uppermost layer is of very low strength and high compressibility, therefore, the most significant compression would take place if the effective stress were to increase throughout the unconsolidated sediments profile as a result of the groundwater extraction, The Bangkok Clay has been divided into the three distinct layers of 'Weather Clay', 'Soft Clay' and 'Stiff Clay' for the purposes of defining its engineering behavior (Moh 1969). The Soft Clay and Weathered Clays are both essentially normally consolidated. The Stiff Clay has relatively high strength and low compressibility; it behaves as a truly over consolidated deposit.

Unconsolidated and semi-consolidated sediments underlying the Bangkok Clay consist of sand, gravel and clay of Pleistocene to Pliocene ages. Aeromagnetic and seismic data covering the Gulf area also indicated an irregular basement by granite ridge and meta-sedimentary fold belts of the peninsular trend (Kelly and Rieb 1971). The basement depth varies from place to place; the maximum depth of 1,830 m was recorded from an oil exploration borehole in the southwest of Bangkok. From a detailed study of electrical log, the upper 600 m of the sediments are subdivided into eight artesian aquifers, separated or partially separated from each other by confining layered or sandy clay beds (Fig. 7.2).

First and the uppermost aquifer immediately overlain by Bangkok clay is Bangkok aquifer (BK). Groundwater in this aquifer is not potable due to high salinity. Most of the groundwater extraction in Bangkok metropolis is from depth of 100–250 m within three productive aquifers Phra Pradaeng (PD), Nakhon Luang

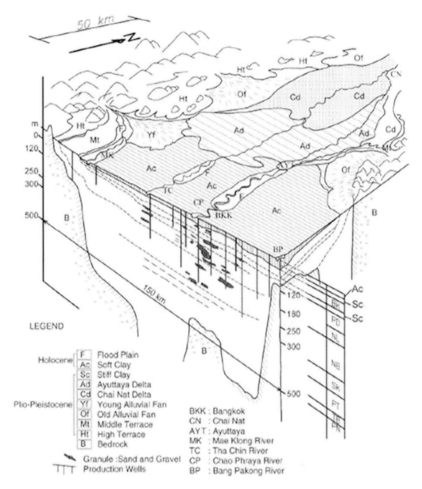

Fig. 7.1 Three-dimensional sketch of landform and stratigraphical section of the Lower Central Plain (Modified after JICA 1995)

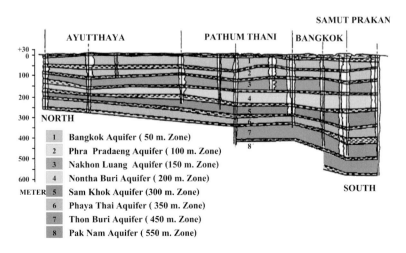

Fig. 7.2 Bangkok aquifer system

(NL) and Nonthaburi (NB). However, there are also many wells tapping the 300 and 350 m zone of Sam khok (SK) and Phaya Thai (PT) in Pathum thani northwest of Bangkok, and groundwater is abstracted for industrial purposes from the 550 m in Pak Nam (PN) aquifer in Samut Prakan south of Bangkok. (Ramnarong 1999).

7.3   Groundwater and Land Subsidence Condition

Groundwater development for public supply in Bangkok began in 1954 with an abstraction of 8,360 m³/d. The daily abstraction for public supply had increased to 450,000 m³ by 1982. Private abstraction was also increasing every year. By 1982, the total daily groundwater abstraction in this area was about 1.4 Mm³/d (Fig. 7.3). The sharply dropped of the total usage between 1985 and 1990 due to the control measures of programme "Mitigation of Groundwater Crisis and Land Subsidence in the Bangkok Metropolis", which became effective in 1983. However the abstraction began to rise again during 1991 onwards due to high economic growth. By 1997, the total groundwater abstraction in the control area of the four provinces; Bangkok, Samut Prakan, Nonthaburi and Pathumthani was 1.67 Mm³/d. There was estimated that groundwater used by licensed users was about 65% of the total abstraction. The remaining 35% is made up of unaccounted by other agencies and unlicensed user.

The initial groundwater levels in Bangkok were very close to ground surface and some wells were said to be artesian. In those days, groundwater is believed to have been pumped from the shallowest good water quality aquifer (PD aquifer) as the overlying Bangkok aquifer (BK aquifer) produced brackish to saline water.

At the early stage of groundwater development for groundwater supply in 1959, water level ranged from 4 to 5 m below surface in eastern Bangkok to 12 m in

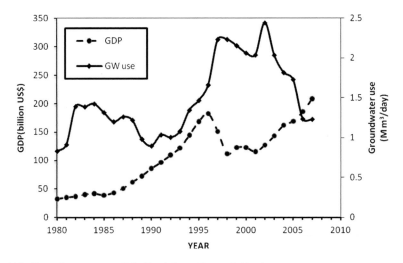

Fig. 7.3 Groundwater use and Thailand Gross Domestic Product

Central Bangkok. After 1967, heavy use of groundwater was observed in the eastern part of Bangkok. By 1967, the lowest water level in NL aquifer in central Bangkok and the eastern suburbs was 30 m below surface. Annual rates of water level decline in NL aquifer during 1969–1974 were 3.6 m (0.7 m/year) in the eastern part and 1–2 m (0.2–0.4 m/year) in central of Bangkok.

During 1959–1982, the water level in the NL aquifer declined by 38 m in central of Bangkok and 60 m in the eastern suburbs (Fig. 7.4). Since 1983, the control measures on groundwater pumping together with the introduction of groundwater tariff in 1985, had a marked effect on the groundwater use in Bangkok; the consequent decrease in groundwater withdrawal produced rapid recovery of water level in the three aquifers (PD, NL, and NB). In central Bangkok, the public water supply produced from surface water sources replace much of groundwater uses, resulting in continuously rise in groundwater level. Most of newly developed industries moved out of the inner zone of Bangkok to its perimeter and the surrounding provinces according to the regulation of Bangkok City Planning. Although the groundwater crisis in central Bangkok has been improved, the cones of depression have developed in the new areas outskirts of Bangkok both in the east province (Samut Prakan) and the west province (Samut Sakhon) which are the areas of extensive industry.

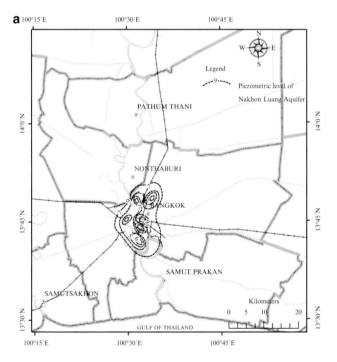

Fig. 7.4 Map showing water level in NL Aquifer from 1959–2005 (**a**) Water level at NL Aquifer in 1959

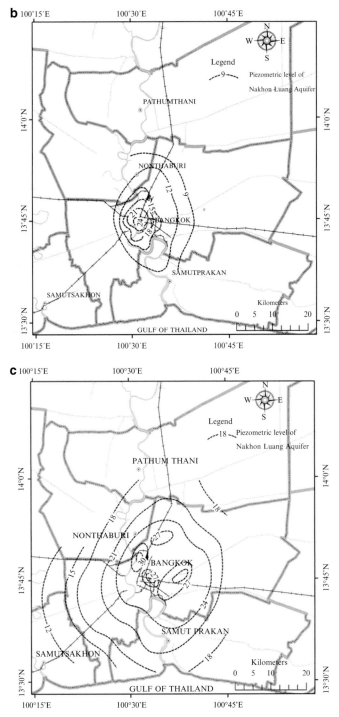

Fig. 7.4 (continued) (**b**) Water level at NL Aquifer in 1969. (**c**) Water level at NL Aquifer in 1974

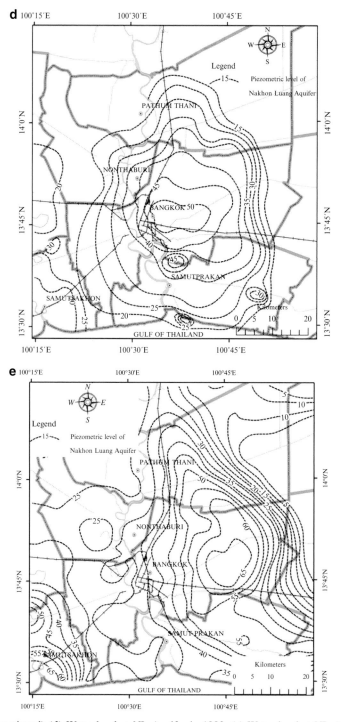

Fig. 7.4 (continued) (**d**) Water level at NL Aquifer in 1982. (**e**) Water level at NL Aquifer in 1996

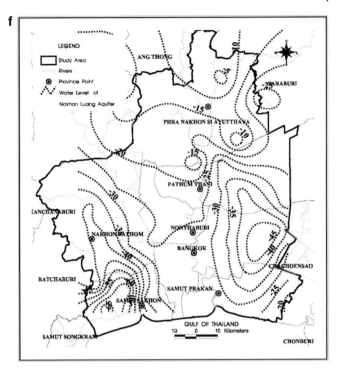

Fig. 7.4 (continued) (**f**) Water level at NL Aquifer in 2005

Land subsidence event can be divides in to three events, early stage, mid-stage, and recent stage. The early stage (1978–1981), land subsidence was over 10 cm/ year in the central Bangkok, and 5–10 cm/year in eastern suburbs. Soil compression during this period in the top 50 m and in the deeper zone, 50–220 m depth, contributed 40% and 60% to the total surface subsidence respectively (AIT 1982). Leveling in 1982 by the Royal Thai Survey Department (RTSD) indicated that the lowest elevation of Bangkok was 4 cm below sea level at a subsidence station in Ramkhamhaeng University.

During the mid-stage, after remedial measures for controlling the groundwater uses were introduced in 1983, a continuous recover of water level was observed in central Bangkok and eastern suburbs. This has contributed to the decreasing rate of land subsidence. The annual subsidence rate (1989) declined to 2–3 cm/year in central Bangkok and 3–5 cm/year in eastern suburbs (Ramnarong and Buapeng, 1992).

At present, subsidence rate has been stabilized and in some area the recovery also present. The overall area subsidence rate is 1 cm/year. The higher rate of 2 cm/ year still can be found at Samut Prakan in the eastern province and Samut Sakhon in the western Province (Lorphensri and Ladawadee 2007). The total land subsidence during two periods are shown in Fig. 7.5. It shows the maximum land depression during 1978–1988 at 70 cm in eastern of Bangkok and during 1978–2005 at 105 cm in the same area.

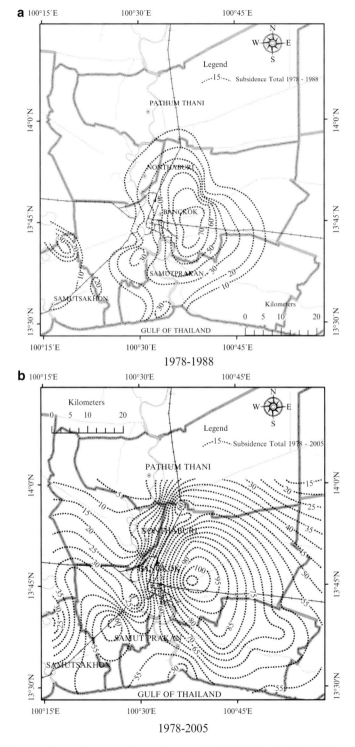

Fig. 7.5 Map showing total land subsidence (in cm). (**a**) 1978–1988. (**b**) 1978–2005

7.4   Control of Groundwater Use

Three methods for controlling the over-exploitation of groundwater in Bangkok and adjacent provinces have been implemented. They are Groundwater Act (1977), Mitigation of Groundwater Crisis and Land Subsidence, Groundwater Tariff and Conservation Fee.

7.4.1   Groundwater Act

Since 1963, the government at that time had a policy to a specific law to regulate usage of groundwater after there was a project surveying groundwater resources in the Northeastern Region to solve drought in 1955. However it was not until 1977 that Thailand enacted the Groundwater Act.

After Thailand introduced its First National Economic and Social Development Plan of 1961–1966, water demands for consumption, agriculture, and industrial sector expanded immensely. Even though the Metropolitan Waterworks Authority (MWA) has been established in 1967 to effectively supply water for household consumption and industrial uses in Bangkok, Nonthaburi, Thonburi and Samut Prakan in order to keep pace with the economic and social development, the MWA was not able to expand its service at the same pace of the economic and social growth. Urban community and industries had to find their own sources of water supply by drilling many groundwater wells to pump substantial amount of water for various uses. The bill on groundwater therefore was retouched for another round of legislative process, but the political situation was not supportive. The bill was then revisited and revised for many times at various stages: the Cabinet, the Council of State as well as the Parliament.

However, when land subsidence in Bangkok Metropolitan was apparent the bill was eventually enacted as the groundwater Act 1977 to control and regulate exploitation of groundwater resources, and to regulate wastewater discharge into the wells. Then, the Groundwater Act 1977 was revised twice, namely in 1992 and 2003.

7.4.1.1   Groundwater Act 1977

Basically the law provides definition of terms concerning groundwater exploitation activities, such as the legal definition of "groundwater", "drilling", "groundwater usage", "well" and other relating terms that needed legal definition. The main concept of this law is that groundwater exploitation is a public matter. Therefore a landowner who wants to drill and exploit groundwater lying under one's own land, one must apply for relevant permits form the Director of the Department of groundwater Resources of from the official authorized to approve on behalf of the Director General.

This concept may be considered as an exemption to the absolute right of property owner as recognized in the Civil and Commercial Code.

In addition the groundwater Act 1977 requires three kinds of permit for different purposes, which are 1-year permit for drilling, 10-year permit for groundwater exploitation, and 5-year permit for discharging water into a well. In this connection, the law sets out rules and conditions for extension of permits, administrative appeal, grounds for refusing permits or extension of one, including certain measures to control and oversee groundwater exploitation. Under the process of permitting for drilling, the provincial groundwater officer can give permitting to the request of the 150 mm well diameter. The requests for larger well diameters are subjected to evaluation by the technical sub-committee under the Groundwater Board of Committee. This technical body provides the technical consultation in regulating both depth and diameter of wells according to the acquired groundwater consumption.

For administrative structure provided by the groundwater Act 1977, there are three institutional parts as follows:

1. The Minister of the Ministry of the Natural Resources and Environment as the highest administrator in the line of public administrator empowered to set out technical rules and regulations on drilling and cease of drilling activities, conservative usage of groundwater, wastewater discharge into well, closure of well, public health and contamination prevention, safety measures for workers and the public, rate of water usage fees (which must be under 1 Baht/m^3, reduction or exemption of fees for users in certain groundwater areas, including other matters concerned. It should be noted here as well that the Minister also has the power to designate areas, critical groundwater areas, and no pumpage areas.

2. Groundwater Board of Committee serving as technical consultant to provide comments or advices to the Minister in designing implementing rules and regulations, including in other relater matters under the law. The Board of Committee may also provide comments or advices to the Director on concerning issues under the law. The members of the board are listed as follows; the Director General of the Department of Groundwater Resources (Chairperson), the Director General of the Department of Public Work, the Director General of the Public Health Department, the Governor of the Metropolitan Waterworks Authority, the Governor of the Provincial Waterworks Authority, the Chairman of The Federation of Thai Industries and another two members which are proposed by the Minister of the Ministry of Natural Resources and Environment.

3. Director General of the Department of Groundwater Resources supervising administrative and managerial matters within the Department of Groundwater Resources to be in accordance with the laws concerning, including the Groundwater Act 1977 and concerning public administration laws. The most important authority of the Director General is granting permit to exploit groundwater and withdrawing the permit. Furthermore, the Director General also has the power to amend the permit if he or she finds any groundwater exploitation dangerous to the environment of respective groundwater exploitation dangerous to the environment of respective groundwater area.

The second revision of Groundwater Act in 1992 was mainly to designate no pumpage area in order to control water quality, to prevent endangerment or deterioration to aquifer, to protect natural resource and environment, to protect public health or properties, or to avoid land subsidence. In addition, the criminal sanction for pumping of groundwater in the no pumpage area and for pumping without the permit, including a criminal procedure for the court to order the offender to restore the well back to its condition before the violation occurred. The implement of "no pumpage area" were aimed at the declared "Critical Zone" which were Bangkok and surrounding six provinces. The method of implementation was that whenever the old permit was expired, the extension was automatically declined. Moreover, the new well would not be permitted. However, this practice was able to be implemented only 2 years, and later on in 1994, groundwater was extremely needed for sustaining the expanding of economic growth. Therefore, the implementation of "no pumpage area" was not strict.

The third revision of Groundwater Act in 2003 was to set up "Groundwater Development Fund" within the Department Of Groundwater Resources to fund study and research on conservation of groundwater and the environment. The whole amount of Groundwater Conservation Tax became the source of Groundwater Development Fund. The fund if to be managed by a board of Board of Executive Committee comprised of the Director General of Department of Groundwater Resources as the Chairperson, representatives from the Bureau of the Budget, the Office of National Economic and Social Development Board, the Comptroller General's Department, the Office of Natural Resources and Environmental Policy and Planning, the Office of Industrial Economic, Department of Water Resources, and the Federation of Thai Industries. The Director General of Department of Groundwater Resources is to appoint Director or an official of equal position within the Department of Groundwater Resources to serve as member and as the Secretary of the board.

7.4.1.2 Groundwater Tariff and Conservation Tax

Groundwater conservation strategy in curbing water demand growth takes a multi-prong approach through pricing, mandatory water conservation. Pricing of water is an important and effective mechanism in encouraging users to conserve water. Groundwater should be treated as an economic good. The water is priced not only to recover the full cost of groundwater management, but also to reflect the scarcity of this precious resource. The process of restructuring of groundwater tariffs and groundwater conservation tax were covered over a 4-year period, starting in 1997, to reflect the strategic importance and environmental impact. The relationship of water level, land subsidence and chronological of mitigation measures were shown in Fig. 7.6.

Groundwater Tariff was first implemented in 1984 in the six provinces of Bangkok and vicinity, where 1.0 Baht/m³ was charged. By 1994, the charge was

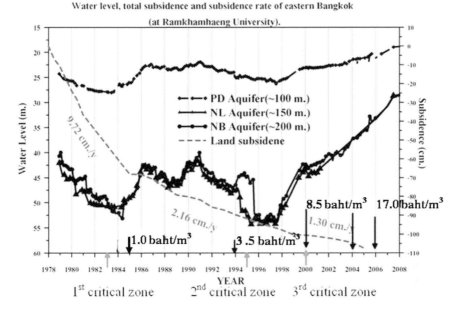

Fig. 7.6 Shows relationship of water level, land subsidence and chronological of mitigation measures. The critical zones were declared and revised in 1983, 1995 and 2000. Total groundwater charge started in 1984 at 1.0 Baht/m³ and reached 17.0 Baht/m³ in 2006

increased to 3.5 Baht/m³, and the government began to charge for groundwater use in the whole country. Between 2000 and 2004, groundwater tariff was gradually increased in the Critical Zone from 3.5 to 8.5 Baht/m³.

In 2003, amendment of the Groundwater Act has recently imposed the Groundwater Conservation Charge for all groundwater users in Critical Zone. Starting 1.0 Baht/m³ in 2004, the charge is set to increase to 8.5 Baht/m³ in 2006. Because of the institution of the Charge, the total cost of groundwater use in the Critical Zone has become relatively high, which has helped in limiting the exploitation of groundwater in the area. Total groundwater charges is expected to increase from 9.5 Baht/m³ in 2004, to 12.50 Baht/m³ by 2005 and to 17 Baht/m³ by 2006 and beyond, which is deterring groundwater users in the area, especially those using large amounts such as industries, from using groundwater for their water supply.

7.4.1.3   Regulatory Measures

The series of regulatory measures were implemented in order to stop the declining water levels to slow the rate of land subsidence. In order to control groundwater use and to mitigate the land subsidence problem, the area which associated

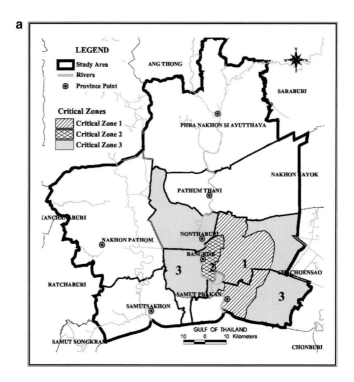

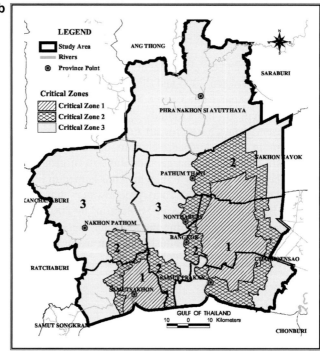

Fig. 7.7 (**a**) Critical Zone declared in 1983, only four provinces were included. The sub-zone 1 referred to area of subsidence rate greater than 10 cm/year, sub-zone 2 referred to subsidence

with land subsidence and groundwater depletion were designated as the "Critical Zone" (Fig. 7.7). The Critical Zone was subjected to strictly control over private and public groundwater users. The different degrees of measures were applied to different degrees of land subsidence (sub-zone). The area of adequate public water supply will not be allowed to put up a new well. Moreover, the requests for groundwater uses by the private sector are critically assessed before any permit is granted. The installing of well meters was enforced in 1985 in support of the charging that the government started to levy from private uses at that time. To promote groundwater and environmental quality conservation, standards for groundwater for drinking purpose were established through the Groundwater Act. In addition, groundwater quality standards for conservation of environmental quality were issued through the Environmental Quality Promotion and Protection Act, 2000.

7.5 Conclusions

The remedial measures for Mitigation of the Groundwater Crisis and Land Subsidence in Bangkok and adjacent provinces were regarded as successful during the first 8 years of their implementation from 1983 to 1990. Recovery of the groundwater level in this period mainly resulted from the abandonment of public supply wells in central Bangkok and the introduction of groundwater tariff that led to a decrease in total abstraction. Since 1991, the total groundwater use has increased each year and seems to be out of control. MWA plans to stop using groundwater for public water supply by 2001 helped decrease the total abstraction.

Groundwater Conservation Strategy by implementation of groundwater tariff and groundwater conservation tax has proven to be the most successful in controlling of groundwater usage and promoting public awareness of important of groundwater to the environment.

Fig. 7.7 (continued) rate of 5–10 cm/year and sub-zone 3 referred to subsidence rate less than 5 cm/year. (**b**) Critical Zone declared in 1995, the total of seven provinces was included. The sub-zone 1 referred to area of subsidence rate greater than 3 cm/year and groundwater level decline rate more than 3 m/year, sub-zone 2 referred to subsidence rate of 1–3 cm/year and groundwater level decline rate of 2–3 m/year and sub-zone 3 referred to subsidence rate less than 1 cm/year and groundwater level decline rate less than 2 m/year. From 2001 up to present, there are no declared critical sub-zones (1, 2, 3). The whole seven provinces were subjected to the same regulations and measures

References

Asian Institute of Technology (AIT) (1972) Effects of deep-well pumping on land subsidence and groundwater resources development in the Bangkok area

Asian Institute of Technology (AIT) (1982) Investigation of land subsidence caused by deep well pumping in Bangkok area, comprehensive report of Asian Institute of Technology, NEB. Pub. 1982–2002

Department of Groundwater Resources (DGR) (2008) Legal development and improvement to conserve groundwater resources. Final report, the environmental law center-Thailand Foundation

Japan International Cooperation Agency (JICA) (1995) The study on management of groundwater and land subsidence in the Bangkok metropolis area and its vicinity. Report submitted to Department of Mineral Resources and Public Works Department, Kingdom of Thailand

Kelly GE, Rieb SL (1971) Tectonic features of the Gulf of Thailland Basin (unpublished map, scale 1:5,000,000)

Moh ZC (1969) Strength and compressibility of soft Bangkok clay, research report no. 7. Asian Institute of Technology, Bangkok

Lorphensri O, Ladawadee A (2007) Report to the Prime Minister office: groundwater and land subsidence situation in Bangkok and the vicinity

Ramnarong V, Buapeng S (1992) Groundwater resources of Bangkok and its vicinity; impact and management. In: Proceeding of a national conference on "Geologic resources of Thailand: potential for future development", Bangkok, Thailand, vol 2, pp 172–184

Ramnarong V (1999) Evaluation of groundwater management in Bangkok: positive and negative. In: Chilton (ed) Groundwater in the urban environment: selected city profiles. Balkema, Rotterdam

Royal Irrigation Department (RID) (2000) Final report on Chao Praya Basin water management project, final report. PAL Consultants, Co., Ltd, Bangkok

Part III
Groundwater Contamination
and Loads to the Ocean

Chapter 8
Detecting Groundwater Inputs into Bangkok Canals Via Radon and Thoron Measurements

Supitcha Chanyotha, Makoto Taniguchi, and William C. Burnett

Abstract Naturally-occurring radon (^{222}Rn) is very concentrated in groundwater relative to surface waters and thus serves as an effective groundwater discharge tracer. Conductivity is also typically present in groundwaters at different levels than associated surface waters and thus may also be used as a tracer of interactions between these water masses. Previous studies by our group using radon and conductivity as groundwater tracers suggested that there is shallow groundwater seeping into the man-made canals ("klongs") around Bangkok. Furthermore, the groundwater was shown to be an important pathway of nutrient contamination to the surface waters. In the present study, we have re-examined some of the same canals and added thoron (^{220}Rn) measurements in order to evaluate if this would provide more site-specific information.

Thoron is a member of the natural ^{232}Th decay chain, has exactly the same chemical properties as radon, but has a much shorter half-life (56 s) than radon (3.84 days). Because of its rapid decay, if one detects thoron in the environment, there must be a source nearby. Thus, thoron is potentially an excellent prospecting tool. In the case of measurements in natural waters, sources of thoron (as radon) could indicate groundwater seeps. During our surveys in the canals of Bangkok, we did successfully measure thoron and its distribution was more variable than that of radon, suggesting that seepage into the canals is not uniform. Areas of higher ground elevation, often in areas where Thai temples are located, were particularly high in thoron.

S. Chanyotha
Department of Nuclear Technology, Faculty of Engineering,
Chulalongkorn University, Bangkok 10330, Thailand
e-mail: supitcha.c@chula.ac.th

M. Taniguchi
Research Institute for Humanity and Nature, 457-4 Motoyama, Kamigamo,
Kita-ku, Kyoto 603-8047, Japan
e-mail: makoto@chikyu.ac.jp

W.C. Burnett (✉)
Department of Earth, Ocean and Atmospheric Sciences, Florida State University,
Tallahassee, FL 32306, USA
e-mail: wburnett@fsu.edu

M. Taniguchi (ed.), *Groundwater and Subsurface Environments: Human Impacts in Asian Coastal Cities*, DOI 10.1007/978-4-431-53904-9_8, © Springer 2011

8.1 Introduction

As part of the "Human Impacts on Urban Subsurface Environments" project (Research Institute for Humanity and Nature), we have been investigating how groundwater becomes an important vector for distributing contamination to surface environments. While groundwater has become an increasingly important aspect of human life, its role as part of the urban environment has not as yet been fully evaluated. This is especially true in Asian coastal cities where population numbers and density have expanded very rapidly and uses of the subsurface environment have increased correspondingly.

An important aspect of the subsurface environment, and the one specifically addressed here, concerns material (contaminant) transport to surface waters. Research over the last few years has shown that direct groundwater discharge to the coastal zone is a significant water and material pathway from land to sea (Moore 1996, 1999; Taniguchi et al. 2002; Slomp and Van Cappellen 2004; Burnett et al. 2003, 2006). While coastal scientists now recognize that groundwater can often be a major contributor to coastal nutrient budgets, most studies to date have been performed in rural, in many cases, pristine environments. This is an understandable desire to deal with "natural" environmental systems. However, with current trends towards global urbanization, it now seems prudent to turn our attention to evaluating such impacts in major urban areas.

Bangkok, originally Khrung Thep ("City of Angels"), is the capital city and most important port of Thailand. After Burmese invaders destroyed the former capital of Ayutthaya in 1767, a temporary capital was established 30 km downstream at Thonburi. Fifteen years later, King Rama I decided to move his palace across to the eastern side of the river where it still stands. His decision to isolate the royal palace resulted in the construction of the first of many canals (called "klongs" locally) that later ran throughout the city. As the economic center of Thailand, Bangkok has an extremely high population density (total population ~10 million). Although it is a major port, it is located some 40 km upstream from the Gulf of Thailand on the Chao Phraya River ("The River of Kings," Fig. 8.1). Beginning in the mid-nineteenth century, roads were built to facilitate land travel, but the river remained the principal artery of communication and the man-made canals served as smaller streets leading into residential districts. Many canals continue to serve as main transport routes to this day.

We conducted a radioisotope and nutrient survey of the Chao Phraya River and Upper Gulf of Thailand in 2004 which was intended to seek out areas of active groundwater discharge to surface waters and to estimate the biogeochemical significance of that discharge to the Upper Gulf of Thailand (Dulaiova et al. 2006; Burnett et al. 2007a). We used continuous ^{222}Rn ("radon") measurements (see below) and collected discrete water samples for measurement of radium isotopes as our main groundwater tracers in that earlier investigation. Samples for measurement of inorganic and organic nutrients and dissolved organic carbon (DOC) were collected at the same stations as the radium samples. The ^{222}Rn levels everywhere in the river were far above the levels that could be supported by its parent, ^{226}Ra (Fig. 8.2). Dulaiova et al. (2006) also showed that these radon levels are higher than

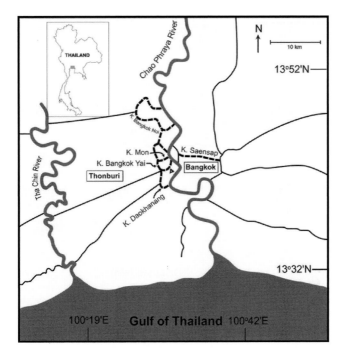

Fig. 8.1 Index map showing the Chao Phraya River and some of the major canals ("klongs," shown as *solid black lines*) that are found in the Bangkok area. We present data here from the K. Bangkok Yai – K. Bangkok Noi series of canals on the Thonburi side of the river (shown as *dashed lines*). K. Saensap, on the east side of the river, was surveyed in February 2007 and the results were reported in Burnett et al. (2009)

what could be expected from diffusion from bottom sediments. Thus, an additional source was required to explain these data.

During this initial study we observed spikes in the radon data that appeared to correspond to locations where major klongs enter the river. This prompted us to return at a later date (2006) and investigate these klongs further. We showed in that more recent study that the canals do receive substantial amounts of shallow groundwater with elevated nutrient concentrations, at least during the rainy season (Burnett et al. 2009). The klongs are thus important pathways for nutrients between the subterranean urban environment around Bangkok and the Chao Phraya River and ultimately to the Gulf of Thailand.

A summary of the 4-day survey performed in 2006 shows that radon activities ranged from values close to those seen in the river (~2,500 dpm/m^3) to over 30,000 dpm/m^3 in parts of the canals (Fig. 8.3). Since the nutrient samples were collected less frequently than the radon measurements, we synchronized the radon data (15-min intervals) to the 27 stations where nutrient samples were collected by averaging the continuous radon measurements centered around the times of nutrient sample collection. We noticed during our surveys that both the conductivity and radon readings tended to show higher values when we passed by one of the many temples ("Wats") that are found on the canals around Bangkok. The smoothed radon data in

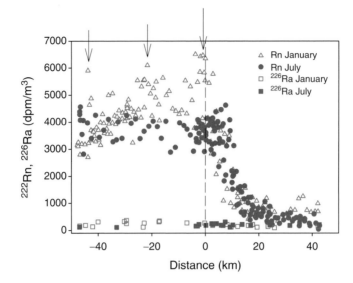

Fig. 8.2 Radon-222 and ^{226}Ra measurements in the Chao Phraya River and Upper Gulf of Thailand from samples collected in January and July 2004. The *dashed vertical line* at 0 km marks the approximate mouth of the river with positive values on the horizontal axis representing distances out to sea and negative values landward. Bangkok is located about 40 km upstream from the Gulf. The spikes in the radon data shown with *arrows* are adjacent to major klongs that enter the river. Data taken from Dulaiova et al. (2006)

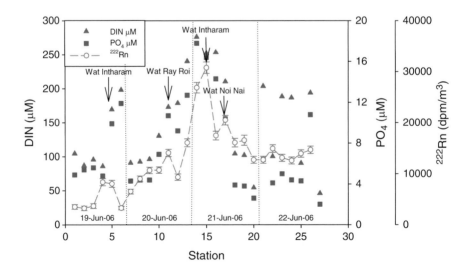

Fig. 8.3 Radon and nutrient (dissolved inorganic nitrogen, DIN, and PO$_4$) results from June 2006 surveys in the K. Bangkok Yai – K. Bangkok Noi canals. Many of the areas high in radon, a groundwater tracer, are also high in these inorganic nutrients. Reproduced from Burnett et al. (2009)

Fig. 8.3 do show some clear peaks occurring adjacent to temple locations. The highest radon concentrations by far occurred around Wat Intharam, on K. Bangkok Yai with additional spikes occurring adjacent to other temples. Note that we passed by Wat Intharam two times (June 19 and 21) and while the peak activity levels were different, there is a spike in the data each time. The different radon activities are likely related to secular variations relating to different tidal stages during the two passes (see Burnett et al. 2009 for more details on these earlier studies).

Higher radon and conductivity around temples may indicate higher groundwater seepage in these areas. But why is this the case? We hypothesized that there could be a connection between higher seepage and the sites where these temples were constructed several centuries ago. One possible explanation is that the temples, being very important centers of the Thai culture, were situated in the most attractive positions possible during the developing years in Bangkok's history (seventeenth and eighteenth centuries). Such favored sites would be located on somewhat higher ground and may have firmer, sandier substrates that could influence groundwater seepage from a shallow aquifer. Most of Bangkok is very low-lying and is covered by a relatively impermeable marine deposit known as "Bangkok clay" (Sanford and Buapeng 1996). Taniguchi (pers. comm.) has followed up this possibility by conducting a series of detailed interviews with monks at several temples to learn more about their history and the decisions that were made concerning the choice of construction sites.

While the earlier studies provided fruitful results, we have been looking for a tracer that provides more spatial resolution. "Thoron" (^{220}Rn), another isotope of radon and a member of the ^{232}Th decay chain (Fig. 8.4), may provide that sensitivity.

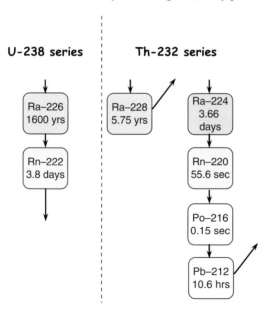

Fig. 8.4 Abbreviated ^{238}U and ^{232}Th decay chains illustrating the relative positions of radon (^{222}Rn) and thoron (^{220}Rn). The remaining natural isotope of radon, actinon (^{219}Rn), is a decay product of ^{223}Ra in the ^{235}U chain

Its very short half-life (56 s) requires that if seen, one must be very close to an active source. With a half-life of less than 1 min, it will essentially decay to zero in approximately 5 min. We recently completed a series of laboratory tests that demonstrated that thoron can, in fact, be measured in natural waters with the same equipment used for our radon measurements (Dimova et al. 2009). Here we present some field data collected in August 2009 from some of the same canals surveyed in 2006 showing how the thoron distributions differ from the radon and may indicate, with a much higher resolution, where active sites of groundwater seepage are occurring.

8.2 Experimental

We performed detailed surveys of several klongs on the western (Thronburi) side of the Chao Phraya River during the period August 25–27, 2009, towards the end of the wet season. The klongs (hereafter abbreviated as K.) surveyed included K. Daokhanang, K. Bangkok Yai, K. Bangkok Noi, and K. Mon (Fig. 8.1). The surveys were conducted from a shallow draft work boat that was chartered for this period. We made continuous measurements of ^{222}Rn, ^{220}Rn, conductivity, temperature, GPS coordinates, and water depth while the boat traveled at a constant and slow speed (4–5 km/h) to enhance the spatial resolution.

Both our ^{222}Rn and ^{220}Rn measurements during the surveys were made with a 3-detector continuous monitor system similar to that described in Burnett et al. (2001) and Dulaiova et al. (2005). A submersible pump delivered near surface water to an air-water exchanger on board the boat while a re-circulating stream of air was pumped through 3 RAD-7 radon detectors (Durridge Co., Inc.) arranged in parallel for measurement. Continuous monitoring of the water-air mixture in the exchanger via a temperature probe allowed for calculation of the radon solubility coefficients and thus conversion from radon-in-air to radon-in-water activities. In order to achieve the maximum sensitivity for thoron, we used a protocol that provided a new reading every 5 min. (in this mode the air pumps will run continuously). We also ran the water pump as fast as possible (~6 L/min) to minimize decay during processing (Dimova et al. 2009). Our radon/thoron mapping system also incorporates integrated global positioning system navigation, depth sounding, in situ specific conductivity and temperature measurements via a Waterloo Scientific CTD. The probe was attached to the harness of the submersible pump and recorded temperature and conductivity at 30-s intervals throughout the surveys.

As previously mentioned, an important advantage of thoron is that its short half-life would ensure that its detection meant that one must be very close to a "source." In the case of groundwater flow into a surface water body, the mean-life of ^{222}Rn is sufficiently long that a radon anomaly could be carried hundreds of meters or even several kilometers away from a discharge point by currents. Thoron, on the other hand, completely decays in about 5 min so a ^{220}Rn anomaly could only be present in surface waters in the immediate vicinity of an active source. Such a source could be related to active groundwater seepage, a submarine spring, high concentrations of [224]Ra in the water, or a high concentration of thorium-bearing minerals (e.g., monazite) in bottom sediments (Burnett et al. 2007b). A few analyses of short-lived radium isotopes in these klongs showed that concentrations are very low (Burnett and Chanyotha, unpublished). While we cannot discount the possibility of elevated thorium-bearing minerals occurring in the canal sediments (generating thoron which could be released by diffusion), we feel that groundwater sources are more likely, especially if a thoron peak is accompanied by anomalies in the radon and/or conductivity.

Another advantage of thoron is its rapid response and decay time, i.e., the ability to see a signal when present and not record a response when absent. Dimova et al. (2009) recently reported results of some detailed laboratory experiments aimed at assessing the optimum settings for radon and thoron using the same equipment as applied here. In order to evaluate the response of the system to radon concentration changes they used a 50-L low-Rn water tank and switched between high radon tap water (groundwater-derived and thus high in radon at >300 dpm/L) to the low level radon reservoir via a two-way valve. They compared their result based on using a 17.5 L/min water flow to an earlier experiment that used a much lower water flow rate (Dulaiova et al. 2005). The results (Fig. 8.5a) showed that at a higher water flow rate there is an improvement in the response time not only when switching from low-to-high concentration but also from high-to-low radon in water, i.e., the radon escapes from the system much faster. The transition time for the previous experimental design was >90 min compared to ~20 min with the high flow modification. So even using the high flow rate modification of the continuous radon measurement system, the "decay time" (time required to re-establish a baseline response) after a spike in the data will cause some smearing out of the spatial resolution when surveying. However, this would not be the case for thoron. Dimova et al. (2009) showed that under the same conditions thoron only requires a few minutes to fully respond to both concentration transition times (Fig. 8.5b).

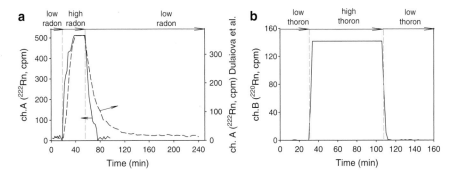

Fig. 8.5 Radon (**a**) and thoron (**b**) response over time during sharp concentration changes using a variation of the standard RAD-7 system. The *dashed line* in (**a**) (right-hand scale) represents results from the standard system with a water flow rate of 5 L/min (Dulaiova et al. 2005). The *solid line* (left-hand scale) is a result with much higher (17.5 L/min) water flow rate. Reproduced from Dimova et al. (2009)

Radon equilibration experiments showed that at water flow rates >5 L/min the equilibrium concentration between water and air is reached very quickly and the only delay in the system response is due to the time for the radioactive ^{222}Rn-^{218}Po equilibrium (~15 min). In the case of ^{220}Rn, the subsequent polonium daughter ^{216}Po has a very short half-life ($T_{1/2} = 0.15$ s) and therefore the pair is almost in instantaneous radioactive equilibrium. This is a significant advantage when prospecting for radon/thoron-enriched sources such as areas of groundwater seepage.

The thoron results reported here were collected in the "B" window (^{216}Po) of each Durridge RAD-7 and were corrected for the approximately 1.5% spectral spillover from the "C" window (^{214}Po) into the ^{216}Po area. We assumed for the purpose of this survey that the efficiency of thoron detection is half that for radon. No attempt was made to account for decay back to the point of sampling. The units for thoron reported are thus arbitrary, uncalibrated units. However, we did make every effort to maintain all conditions (water and air flow rates, etc.) uniform so we feel that the relative values reported for thoron are correct. For the "prospecting" purposes intended here, relative values are sufficient. The radon results are based on the count rate in window "A" (^{218}Po), and are fully calibrated and corrected for air-water temperature differences, etc.

8.3   Results and Discussion

8.3.1   General Trends

Since all of the radon and thoron data were collected with a 5-min time integration and the conductivity data were collected every 30 s, the results needed to be synchronized before making comparisons. We combined and filtered the specific conductivity data so that the mid-point of the integrated results ($n = 11$ for each radon measurement) was the same as the radon midpoint. We then plotted the radon values for each day against the synchronized conductivities (Fig. 8.6).

The radon-conductivity results show much the same pattern as we saw in these canals during the wet season in 2006, i.e., most of the data can be explained by mixing between three end-members: (1) the Chao Phraya River (relatively low radon and low conductivity); (2) an area of high radon and moderate conductivity (typified by some of the data from K. Daokhanang); and (3) an area of moderate radon and high conductivity (K. Bangkok Noi). The approximate end-member values for the river (2,500 dpm/m^3, 350 µS/cm) and K. Bangkok Noi (10,000 dpm/m^3, 800 µS/cm) are almost exactly the same as we measured in 2006 (see Fig. 8.4 in Burnett et al. 2009). The high radon, moderate conductivity end-member (23,000 dpm/m^3, 500 µS/cm), however, is somewhat lower in both radon and conductivity than that seen in June 2006 (~32,000 dpm/m^3, 700 µS/cm). This may be a consequence of surveying a somewhat different series of canals off K. Daokhanang on August 27th than we did in June 2006. In addition, there are known secular changes in these parameters both on time scales of hours (tidal effects),

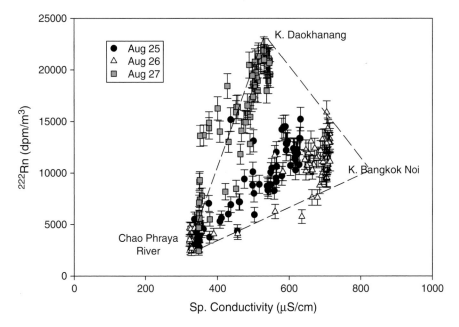

Fig. 8.6 Rn-222 versus specific conductivity for the canal surveys of August 25–27, 2009. Most of the data can be explained by mixing between three end-members as indicated on the diagram

as shown by a time-series experiment we performed in 2006 (see Fig. 8.5 in Burnett et al. 2009), as well as seasonal effects (wet versus dry periods). For example, data collected in February 2007 (dry season) showed very low radon levels in some of these same canals indicating that the groundwater seepage may only be active in these canals during the wet season.

Based on these results, it is clear that there are areas within the klongs that have elevated concentrations of groundwater tracers. While high conductivities in klong waters could be the result of various wastewater discharges in this urban setting, it is unlikely that the types of industrial or domestic effluents present in the Bangkok area would also contain high radon. We feel that the most likely interpretation of the areas that contain both elevated conductivities and high radon is that these areas are characterized by groundwater seepage.

8.3.2   Thoron Measurements

When we plot thoron versus radon (Fig. 8.7) it is clear that their distributions are completely independent of each other. The canals around K. Daokhanang surveyed during day-3 clearly had the highest radon but the highest thoron measurements were seen in K. Bangkok Noi on the 2nd day of the surveying. This may seem surprising at first glance, especially since we propose that they both have a common

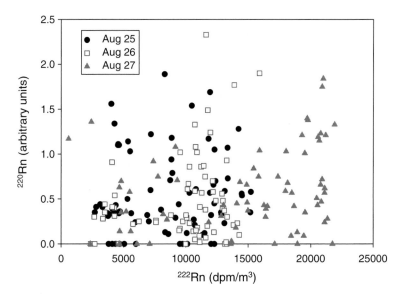

Fig. 8.7 Scatter plot showing activities of ^{220}Rn (thoron, arbitrary units) versus ^{222}Rn (radon, dpm/m^3) for each of the 3-days of the survey

source (groundwater seepage). However, because of the difference in half-lives (3.84 days for radon, 56 s for thoron), this can easily be explained. An average radon atom that enters a canal will persist for something like 5.5 days (mean life of ^{222}Rn $= 1/\lambda$) before it decays or is otherwise removed from the system by current transport or atmospheric evasion. An average thoron atom, on the other hand, has a mean life of only about 80 s. If thoron doesn't emanate to the atmosphere or is otherwise transported away from its entry point, it will decay away to essentially zero in about 5 min. Thus, radon may enter canals, spread out via current action (the canals are tidally influenced), and persist for days resulting in more uniform spatial trends. Thoron, however, will only be measured if we happen to sample water very close to a source. If this source should be groundwater, as we think that it is, a spike in the thoron must indicate we are in very close proximity to a point of groundwater seepage. We will illustrate in the following section the contrasting trends between the radon/conductivity tracers, which behave in a somewhat similar manner, and the thoron distribution.

8.3.3   K. Bangkok Noi Results

The potential usefulness of thoron as a prospecting tool is best illustrated from the data set collected along K. Bangkok Noi, the northernmost of the canals surveyed during this investigation. The route we used, entering at the north and following a counter-clockwise direction towards its confluence with K. Bangkok Yai, was reasonably long (~25 km) and passed areas of varying land-use (mostly a combination of either low-lying natural vegetated areas or residential built-up areas). There is thus the opportunity to see if thoron responds to differences in elevations or other changing characteristics along this path that could influence groundwater seepage.

When we plot radon, thoron, and conductivity as a function of time or distance along the survey route we see that the trends for radon and conductivity are very similar while the thoron distribution is much different (Fig. 8.8). The sharp rise observed in conductivity and radon in the late morning is not only related to probing further into the canal system but also to a tidal change. We began the survey during an incoming tide which was delivering relatively low radon and low conductivity waters into the canal, diluting the higher radon and conductivity found there. Once we were several kilometers inside the canal the tide turned to an outgoing flow, the conductivities and radon quickly increased and then remained at more-or-less high and constant levels. The thoron distribution, on the other hand, shows several distinct peaks along the route. This is likely a reflection of the much faster response/decay time of the thoron compared to radon, i.e., as a qualitative indicator it is much more sensitive. At least four of the thoron peaks (numbered 1–4 in Fig. 8.8) line up with smaller secondary peaks in the radon (but no distinguishable peaks in the specific conductivity). All of these four peaks, as well as some of the others, correspond to temple locations along the survey route. Thus, as first seem in the 2006 surveys, there is some suggestion that areas around temples may have elevated groundwater seepage.

Another way to look at these trends is to plot the results on a GIS base map that has additional land-use information. We have done this for temperature and

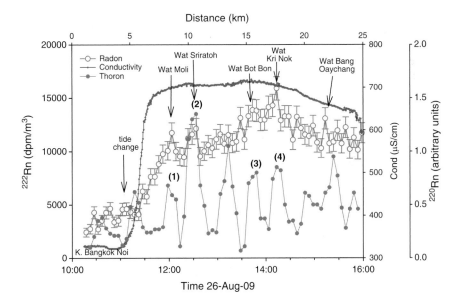

Fig. 8.8 Radon, thoron, and conductivity versus time and distance for the K. Bangkok Noi survey of August 26, 2009. The thoron data is shown as a 3-point moving average. The radon and conductivity data are not smoothed

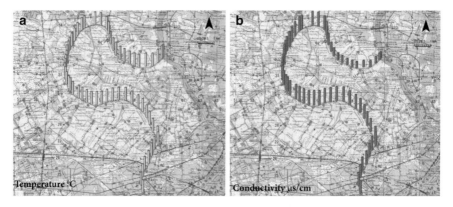

Fig. 8.9 GIS maps showing temperature (*left*; 32°C max.) and specific conductivity (*right*; 637 μS/cm max.) distributions along the survey route for August 26, 2009. Both temperature and conductivity, originally measured at 30-s intervals, have been synchronized to the 5-min radon/thoron measurement cycle. The base map displays general land-use with built-up areas in the *darker color* and natural vegetated areas in the *light background*. Temple locations are shown as *circles*

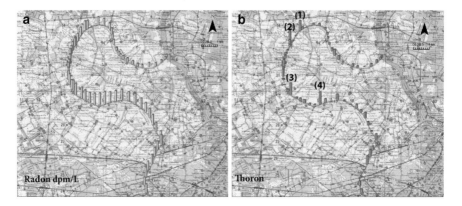

Fig. 8.10 GIS maps showing radon (*left*; 16 dpm/L max) and thoron (*right*, arbitrary units) distributions along the survey route for August 26, 2009. The base map displays general land-use with built-up areas in the *darker color* and natural vegetated areas in the *light background*. Temple locations are shown as *circles*

conductivity (Fig. 8.9) and for radon and thoron (Fig. 8.10) measurements made on the same day as discussed above (August 26, 2009). The base map used provides some land-use information with the darker color representing built-up areas (private houses, temples, schools, etc.) and the lighter color representing natural vegetated areas. When displayed in this manner, there are no discernable trends in the temperature (although some trends due to tidal mixing and solar warming could be seen when the full data set is displayed on an x-y plot). The increase in conductivity as one enters the canal can easily be seen in the conductivity data.

The radon data (Fig. 8.10) shows much the same trend as the conductivity when observed in this way. The smaller secondary peaks that can be seen in Fig. 8.8 cannot easily be detected in this GIS view. The thoron results, however, show clear peaks at different points along the survey path (peaks 1–4 correspond to the same peaks in Fig. 8.8). Most, although not all, of the thoron peaks appear along the built-up areas on K. Bangkok Noi. Of a total of nine thoron peaks that can be identified in Fig. 8.8 (smoothed peaks greater than 0.5 units), seven occur adjacent to temple complexes along the canal. All of these are also within the built-up areas according to the GIS maps. It has been confirmed that in many cases, sand fill was added to the ground as a substrate during the construction of these temples (Taniguchi, pers. comm.). Thus, the "temple-groundwater" relationship may be real as the higher elevations and sand fill would be more conducive to groundwater flow.

8.4   Summary and Recommendations

The results presented here were all collected within a few days and not under the most optimal experimental conditions. Thus, the thoron data should be considered somewhat preliminary although the trends in the data do support earlier findings and are thought to be correct. As Dimova et al. (2009) recently showed, the sensitivity for thoron can be increased significantly by use of a high-speed submersible water pump and an external air pump in addition to the internal pumps of the RAD-7 radon detector. These enhancements were not available to us during these field surveys. In spite of using a system designed more for radon than thoron, we were able to see clear thoron peaks in these canals. A follow-up survey, using a system optimized for thoron, would allow for a firm confirmation of these results. In addition, both the June 2006 and August 2009 studies were performed during wet season periods. Only a single series of measurements is available during the dry season (February 2007) and that indicated much lower groundwater tracers. Additional surveys at different points in the monsoonal season are needed to confirm if there is indeed a seasonal pattern to the groundwater discharge into the canals of Bangkok.

Acknowledgments The research presented here is a contribution to the "Human Impacts on Urban Subsurface Environments" Project of the Research Institute for Humanity and Nature (RIHN) in Kyoto, Japan. This investigation was conducted while W. Burnett was a visiting scientist at RIHN during the summer of 2009. The authors acknowledge RIHN for the financial, intellectual, and personal support that made this investigation possible. We also thank the following students from the Department of Nuclear Technology, Chulalongkorn University, for their excellent assistance during the field work: Ms. Rawiwan Kritsananuwat, Mr. Phongyut Sriploy, Ms. Karnwalee Pangza, and Mr. Jumpot Jamnian.

References

Burnett WC, Kim G, Lane-Smith D (2001) A continuous radon monitor for assessment of radon in coastal ocean waters. J Radioanal Nucl Chem 249:167–172
Burnett WC, Bokuniewicz H, Huettel M, Moore WS, Taniguchi M (2003) Groundwater and porewater inputs to the coastal zone. Biogeochemistry 66:3–33

Burnett WC, Aggarwal PK, Bokuniewicz H, Cable JE, Charette MA, Kontar E, Krupa S, Kulkarni KM, Loveless A, Moore WS, Oberdorfer JA, Oliveira J, Ozyurt N, Povinec P, Privitera AMG, Rajar R, Ramessur RT, Scholten J, Stieglitz T, Taniguchi M, Turner JV (2006) Quantifying submarine groundwater discharge in the coastal zone via multiple methods. Sci Total Environ 367:498–543

Burnett WC, Wattayakorn G, Taniguchi M, Dulaiova H, Sojisuporn P, Rungsupa S, Ishitobi T (2007a) Groundwater-derived nutrient inputs to the Upper Gulf of Thailand. Continental Shelf Res 27:176–190

Burnett WC, Dimova N, Dulaiova H, Lane-Smith D, Parsa B, Szabo Z (2007b) Measuring thoron (^{220}Rn) in natural waters. In: Warwick P (ed) Environmental radiochemical analysis III. Royal Society of Chemistry, RSC Publishing, Cambridge, pp 24–37

Burnett WC, Chanyotha S, Wattayakorn G, Taniguchi M, Umezawa Y, Ishitobi T (2009) Groundwater as a pathway of nutrient contamination in Bangkok, Thailand. Sci Total Environ 407:3198–3207

Dimova N, Burnett WC, Lane-Smith D (2009) Improved automated analysis of radon (^{222}Rn) and thoron (^{220}Rn) in natural waters. Environ Sci Technol 43:8599–8603

Dulaiova H, Peterson R, Burnett WC (2005) A multi-detector continuous monitor for assessment of ^{222}Rn in the coastal ocean. J Radioanal Nucl Chem 263(2):361–365

Dulaiova H, Burnett WC, Wattayakorn G, Sojisuporn P (2006) Are groundwater inputs into river-dominated areas important? The Chao Phraya River – Gulf of Thailand. Limnol Oceanogr 51:2232–2247

Moore WS (1996) Large groundwater inputs to coastal waters revealed by ^{226}Ra enrichments. Nature 380:612–614

Moore WS (1999) The subterranean estuary: a reaction zone of ground water and sea water. Mar Chem 65:111–125

Sanford WE, Buapeng S (1996) Assessment of a groundwater flow model of the Bangkok Basin, Thailand, using carbon-14 based ages and paleohydrology. Hydrogeol J 4:26–40

Slomp CP, Van Cappellen P (2004) Nutrient inputs to the coastal ocean through submarine groundwater discharge: controls and potential impact. J Hydrol 295:64–86

Taniguchi M, Burnett WC, Cable JE, Turner JV (2002) Investigations of submarine groundwater discharge. Hydrol Processes 16:2115–2129

Chapter 9
Subsurface Pollution in Asian Megacities

Shin-ichi Onodera

Abstract To confirm the various subsurface pollutants associated with urbanization, many previous studies were reviewed. Nitrate, trace metal, and chloride subsurface pollutants have considerably increased with urbanization. Megacities, in particular, are highly vulnerable to pollution. Some of the Asian megacities were classified as follows: Jakarta and Manila are in the developing stage (1st stage) with serious surface pollution problems from nitrate and trace metals: Bangkok is in the developed stage (2nd stage) with subsurface pollution from various contaminants: and Seoul and Taipei are in the developed stage with infrastructure for sewage treatment (3rd stage), as are Tokyo and Osaka. The third stage cities may experience potential delayed contaminant discharge through groundwater to rivers and to the sea in the surrounding area. The vulnerability of each city to pollution was determined by the intensities of surface and subsurface pollution. These intensities are controlled by human impact as well as the natural environment. For example, the emission and load of pollutants increased with the population and surface pollution occurred in the city's first stage of development. In the next stage, subsurface pollution occurred with the transport of surface pollutants to groundwater. In addition, groundwater abstraction affected the intrusion of surface pollution to deep groundwater. On the other hand, contamination and attenuation processes related to groundwater flow conditions were controlled by the natural environmental factors, such as the topography, geology, watershed area, and natural recharge or climate.

9.1 Introduction

Subsurface pollution such as soil, sediment, and groundwater pollution is generally caused by human activities. For example, groundwater pollution has been reported in megacities, as well as in agricultural areas (Burt et al. 1993; Environment Agency

S.-i. Onodera (✉)
Graduate School of Integrated Arts and Sciences, Hiroshima University,
1-7-1 Kagamiyama, Higashi-Hiroshima, Hiroshima 739-8521, Japan
e-mail: sonodera@hiroshima-u.ac.jp

M. Taniguchi (ed.), *Groundwater and Subsurface Environments: Human Impacts in Asian Coastal Cities*, DOI 10.1007/978-4-431-53904-9_9, © Springer 2011

Fig. 9.1 Wells dug near dump sites in Jakarta (**a**) and Manila (**b**)

Japan 1996; Appelo and Postma 2005). In megacities, there are many sources of contamination involved in the mass cycle: (1) large quantities of materials are generally accumulated and consumed; (2) large quantities of various types of waste are produced, burned, and discarded; and (3) large amounts of contaminants are consequently discharged to rivers, groundwater, and the ocean. Therefore, both air and water pollution occurs around these cities (World Bank 1997; Tsunekawa 1998). In addition, such contaminated water is often consumed by the population of megacities in developing countries. Figure 9.1a, b show the condition of wells located near dump sites in Jakarta and Manila, respectively. It is necessary to determine the water quality and show the risk of using this water in such cities.

Generally, the transport of contaminants is influenced by biogeochemical processes (Appelo and Postma 2005), as well as by groundwater flow. Slomp and Van Cappellen (2004) noted the effects of redox processes on nutrient transport in groundwater in coastal areas. For example, nitrate (NO_3^-) is denitrified under anaerobic conditions, and this reaction often occurs in groundwater discharge areas (Howard 1985; Hinkle et al. 2001; Böhlke et al. 2007). The function of the natural attenuation of contaminants through processes such as denitrification should also be evaluated along with the contamination by trace metals (Gandy et al. 2007).

The aim of this chapter is to confirm the various types of subsurface pollution associated with urbanization, and to summarize the relationship between the development stage of a city and the level of pollution.

9.2 Effect of Urbanization on Groundwater Conditions

The population of Asia has increased rapidly over the last three decades. There were six megacities in Asia having populations of more than ten million people in 1995 (UN 1999; Jiang et al. 2001), then the number reached 14 in 2010 (Brinkhoff 2010). Tokyo and Osaka became megacities in the 1960s; Seoul, Shanghai, and Mumbai in the 1980s; Beijing, Guangzhou, Delhi, and Kolkata in the 1990s; and Manila, Karachi, Jakarta, Dhaka, and Tehran in the 2000s. The populations of several other

Asian cities will also approach ten million up to 2020, e.g., Shenzhen, Wuhan, Bangkok, Lahore, Tianjin, and Bengaluru. Such drastic increase in population results in the high consumption of resources such as water, energy, and food.

In Tokyo and Osaka, groundwater rather than surface water was used as the main water resource during the period of rapid population increase before 1970 (Environment Agency Japan 1969). The surface water was not suitable for drinking because of contamination and insufficient volume (Foster 2001). Consequently, groundwater resources in Tokyo and Osaka were degraded (Environment Agency Japan 1996). After this period, however, groundwater use was restricted because of pollution and the decline in the groundwater level.

For sustainable use of groundwater in developing Asian megacities such as Jakarta, Manila, and Bangkok, it will be important to conserve groundwater quality and quantity based on the experience of developed megacities such as Tokyo and Osaka. However, the process of groundwater degradation, including depletion and contamination, has not been sufficiently examined in these developing megacities. In addition, it is necessary to clarify the variation in groundwater flow and solute transport as well as contaminant inputs and their sources.

9.3 Developing Stage of Asian Cities and Subsurface Pollution

9.3.1 A Case Study of Jakarta City

Jakarta is located on the lowlands of the northern coast of West Java province in Indonesia (6°15′ S, 106°50′ E) and covers an area of approximately 650 km^2. Jakarta has a humid tropical climate; the average annual temperature is 27°C, and the yearly rainfall is 2,000 mm due to the influence of the monsoons. Jakarta has a population of approximately 12 million people. Land subsidence is one of the serious environmental problems in Jakarta, with estimated subsidence rates of 1–10 cm year^{-1} (Abidin et al. 2007).

Generally, air pollution is one of the factors affecting groundwater pollution. Burt et al. (1993) showed that nitrate originating from air pollution is transported to groundwater. Table 9.1 shows the condition of air pollution in some Asian megacities (Jiang et al. 2001). The mean annual total suspended particulate (TSP) in Osaka and Tokyo are one order of magnitude lower than those in Jakarta, Manila, and Mumbai, and the mean annual SO$_2$ concentrations in Osaka and Tokyo are half of those in Manila and Mumbai. These results suggest that the pollution load to groundwater by rainfall and dry fall in Jakarta is larger than that in Osaka. The population in each of these cities is approximately more than ten million. However, the population growth situation is different. The population of Osaka is gradually decreasing; on the other hand, Jakarta's population continues to increase. The growing population of big cities intensifies the problem of air and groundwater pollution.

Table 9.2 shows the condition of water pollution and pollution loads in various countries surrounding the South China Sea (Jiang et al. 2001). The biochemical

Table 9.1 Air pollution in Asian megacities (modified from Jiang et al. 2001)

Megacity	Population (million)	Mean annual TSP (μg m^{-3})	Mean annual SO$_2$ (μg m^{-3})
Osaka	10.61	43 (1993)	19 (1994)
Tokyo	26.96	49 (1993)	18 (1995)
Jakarta	8.62	271 (1990)	No data
Manila	9.29	200 (1995)	33 (1993)
Mumbai	15.14	240 (1994)	33 (1994)

Table 9.2 Population of watersheds (SCS population) and dissolved loads to the South China Sea (SCS) (modified from Jiang et al. 2001)

Country	SCS population (million)	BOD (10^3 ton year^{-1})	TN (ton year^{-1})
Cambodia	1.98	36.2	361
China	59.69	1,089.4	8,716
Indonesia	105.22	1,920.2	592×10^6
Malaysia	10.34	188.6	
Philippines	23.63	431.3	
Thailand	37.14	677.8	>3,037
Viet Nam	75.12	5,714.5	21,987

oxygen demand (BOD) and total nitrogen (TN) load in Indonesia is 2 times and 1,000 times that in China, respectively. In addition, the BOD load in Viet Nam is three times that in Indonesia. It is necessary to control the pollution load in such countries to conserve the water and coastal environment.

9.3.2 A Case Study of Osaka City

Figure 9.2 shows Osaka city, Japan, and the surrounding areas. The area shown in the darker shade is Osaka city, and the pale colored area is the Osaka metropolitan district. The latter includes Kobe city and Kyoto city. The population of the Osaka metropolitan district is over ten million. This area is characterized by a relatively small suburban area.

Figure 9.3a shows the variation in population and the industrial production index of Osaka prefecture, both of which increased significantly from the 1920s to the 1970s. The population was less than three million in 1945, but grew to more than eight million in the 1970s. The urban area expanded from the city center to the surrounding area, shown by the medium shade color in Fig. 9.2, as the population increased. Consequently, the suburban area decreased in these areas. Since the 1970s, the prefecture population has increased by slightly less than one million in 30 years. However, the industrial production index has increased by 1.5 times of that in 20 years.

Intensive groundwater pumping caused the lowering of the groundwater level and land subsidence during 1960s. The groundwater level was less than −25 m below sea level, and the annual land subsidence rate was about 10 cm. However,

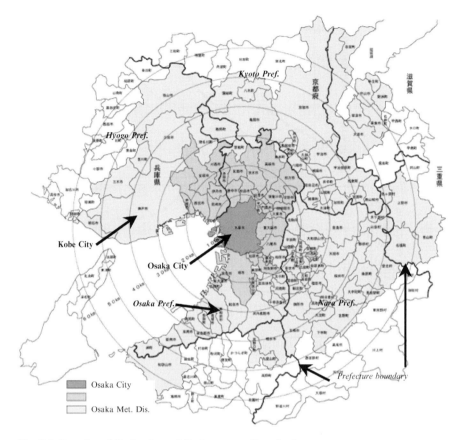

Fig. 9.2 Location of Osaka city and Osaka metropolitan district

due to government regulation of water pumping, the groundwater level recovered after 1970 and the land subsidence also stopped.

Figure 9.4 shows the variation of hydraulic head of coastal deep groundwater over a period of 20 years. The hydraulic head is the groundwater level represented by the altitude. The hydraulic head in deep groundwater has increased, but it remains less than the sea level, that is, seawater intrusion continues to occur and groundwater is not discharged into the ocean.

Figure 9.3b shows the variation in surface chemical oxygen demand (COD) concentration in Osaka Bay and COD load from the river to the sea for the last 75 years (Nakatsuji 1998). The COD concentration and COD load reached their maximum values around 1970. The COD load from the river to the sea was at minimum around 1950, and was approximately constant before 1950. The COD load in the river increased by a factor of 4 in the 20 years from 1950 to 1970. This period coincides with a rapid rise of the population in the area. These results indicate the effect of urbanization on the quality of river water and seawater.

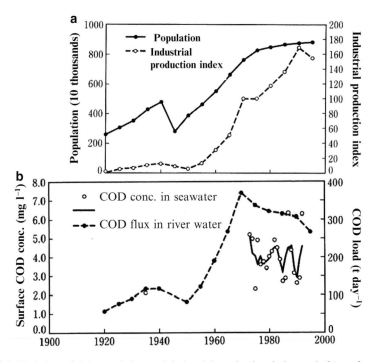

Fig. 9.3 Variation of (**a**) population and industrial production index and (**b**) surface COD concentration in Osaka bay and COD flux from river (Nakatsuji 1998)

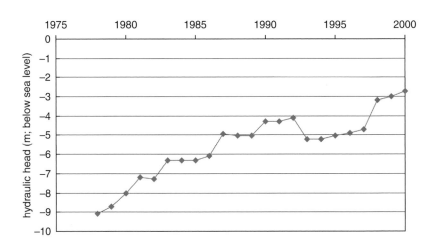

Fig. 9.4 Variation in hydraulic head of deep groundwater in Osaka

Since the 1970s, the population has gradually increased and the industrial production index has also increased; however, the COD concentration and load have decreased. This trend can be attributed to the development of sewage treatment systems. However, even though the river water pollution decreased, the impact of the pollution that

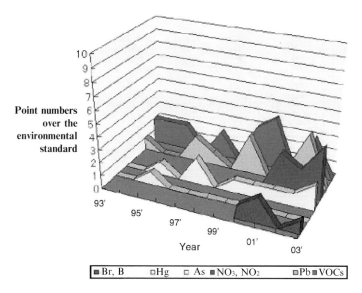

Fig. 9.5 Variation in contaminant species of groundwater from 1993 to 2003 in Osaka

occurred before the 1970s would still be reflected in the subsurface environment. Burt et al. (1993) introduced the example of sluggish transport of contaminants accumulated for 30 years in an unsaturated zone to the saturated zone in upland, England. Therefore, we need to consider the groundwater contamination that will occur following the peak of river water pollution. In fact, the COD in the seawater of Osaka Bay shows a relatively constant trend with some variations (Fig. 9.3b). This suggests the effect of contaminated groundwater discharge as well as the buffer effect of the large volume of the bay. Since the 1990s, the industrial production index has decreased with the removal of industries from the area (Fig. 9.3a). Consequently, the COD concentration in the seawater and the load in the river have also decreased.

Figure 9.5 shows the condition of groundwater contamination in Osaka prefecture from 1993 to 2003 (Environment Council, Osaka Prefectural Government 2004). The contaminant species in this figure include bromide (Br), boron (B), mercury (Hg), arsenic (As), nitrate (NO_3), lead (Pb), and volatile organic compounds (VOCs). This result indicates that various contaminants have been detected in groundwater in the last decade.

9.3.3   Properties of Pollution in Cities at Various Stages of Development

Based on the example of Osaka, I attempt to discuss the pollution characteristics of cities in various stages of development. In the 1950s, some pollution and environmental damage by urbanization was beginning to appear on a local level such as the river in Osaka city. This period is the first stage of water pollution due to accelerated

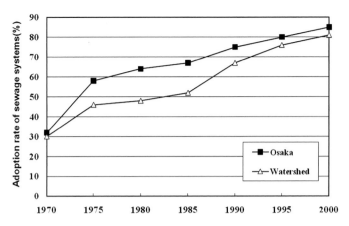

Fig. 9.6 Variation in adoption ratio of sewage systems from 1970 to 2000 in Osaka city and a watershed of a river flowing in Osaka city

economic growth and population increase in the megacity. At this stage, the main contaminant was composed of dissolved nitrogen from domestic and agricultural wastewater, and heavy metals originating in industrial activity. In the 1960s, Japanese megacities had experienced of the most severe contamination in rivers and inland seas from human sewage and industrial waste.

To prevent severe water pollution, the basic laws for Environmental Pollution Control and Water Pollution Control were introduced in 1967 and 1970, respectively. Since the 1970s, pollution of the surface and river environments began to decrease as the accelerated population growth stopped. This period is the second stage. The land around the city changed from agricultural land to residential land, and the urban area expanded to the suburban area. This change caused a decrease in the amount of agricultural waste generated, which is the source of high levels of nitrate. Figure 9.6 shows the variation in the adoption ratio of the sewage system from 1970 to 2000 in Osaka city and a watershed of the Yodo River, which flows through Osaka city. The ratio increased from 30% to 65% for a decade in 1970s. The adoption of urban infrastructure such as sewage systems also contributed to the remediation of surface pollution. However, for over 20 years after the late 1970s, we were still faced with groundwater and soil contamination from nitrate, heavy metals, and organic compounds. This indicates that the contaminants had been accumulating in the subsurface since the first stage of the city's development.

The adoption ratio of sewage systems exceeded 80% by 1995 in Osaka city and by 2000 in the Yodo River watershed (Fig. 9.6). In Osaka, the remediation of river pollution had progressed with the development of urban infrastructure. This period is the third stage. The land use has also been changed from industrial to public and residential purposes because of the appreciation of land values. These changes contributed to a further decline in pollution. However, we can still find the subsurface pollution. These subsurface contaminants will be transported to the river and sea with a delay. The distribution of trace metal content in the sediment in various Asian megacities, as indicated by previous studies

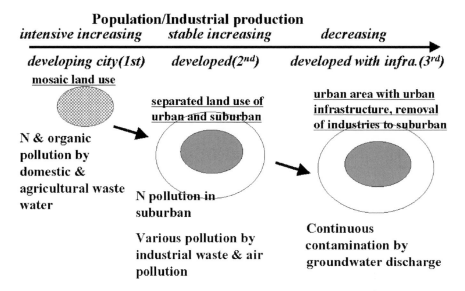

Fig. 9.7 Schematic diagram of conceptual model of relationship between developing stage of megacity and pollution problem

(Williams et al. 2000; Jiang et al. 2001; etc.), shows the change in the pollution properties during the growing stage of a city.

Half of the world's megacities are distributed in Asia in 2010. However, these megacities have different environmental problems in terms of the stage of development or pollution environment. Based on our conceptual model, some megacities can be classified as follows: Jakarta and Manila are in the developing stage (1st stage in Fig. 9.7) and suffer from serious surface pollution problems due to nitrate and trace metals. Bangkok with the low annual population growth is in the developed stage (2nd in Fig. 9.7) and suffers subsurface pollution from various contaminants. Finally, Seoul and Taipei are in the developed stage with infrastructure for sewage treatment (3rd in Fig. 9.7), as are Tokyo and Osaka, and have a potential of delayed contaminant discharge from groundwater to the rivers and sea.

9.4   Nitrate Pollution

Water pollution by nitrate is a problem worldwide, particularly in agricultural regions (Burt et al. 1993). Shindo et al. (2003) reported serious nitrate pollution in China, using the nitrogen cycle model. In general, nitrate in groundwater originates mainly from municipal and industrial waste, air pollution, and leaching from agricultural land. In China, agricultural leaching was a major issue. However, in megacities, there are many other sources.

Spatial variations in dissolved nitrogen in the groundwater of some Asian megacities have been reported by Umezawa et al. (2009a). The results in Manila, Bangkok, and

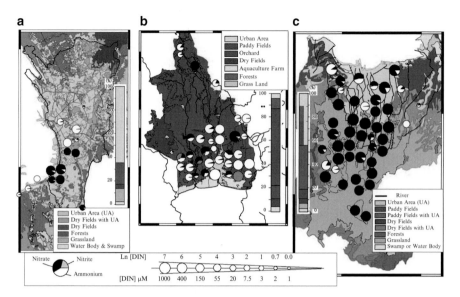

Fig. 9.8 Spatial distribution of dissolved nitrogen concentrations of groundwater and land use in Manila (**a**), Bangkok (**b**), and Jakarta (**c**) (Umezawa et al. 2009a)

Jakarta are shown in Fig. 9.8. The environmental standard concentration is 10 mg l⁻¹ and 714 μM l⁻¹ of nitrate nitrogen, as determined by the US Environmental Protection Agency (USEPA) and Ministry of Environment (ME), Japan. The World Health Organization (WHO) environmental standard concentration is 50 mg l⁻¹ and 806 μM l⁻¹ in nitrate. The points exceeding this value were few, but the concentrations of nitrate were still evident. The species of dissolved nitrogen was mainly ammonium in Bangkok and nitrate in Jakarta. These characteristics suggest that the groundwater of Bangkok is in an anoxic and reductive condition, and that in Jakarta it is in an oxic condition. However, these samples in Bangkok were mainly collected from deep groundwater; therefore, it will be necessary to confirm shallow groundwater pollution according to the well connection with the rivers and canals (Burnett et al. 2009).

In addition, the results showed the nitrogen and oxygen stable isotopic ratio of nitrate in the groundwater (Fig. 9.9). According to Kendall and McDonnell (1998), the various sources of nitrate and the denitrification trend can be estimated, using these isotopic ratios. If both values are around 0 in this diagram, it means the nitrate source is a chemical fertilizer. On the other hand, if it is plotted around (0, 20) and (20, 0), it suggests the nitrate source is from air pollution or municipal waste, respectively. In addition, if both values increase along the flow line, a trend of denitrification process is suggested (Postma et al. 1991; Böhlke and Denver 1995; Tesoriero et al. 2000). The results in Fig. 9.9 indicate that the nitrate sources in the groundwater of these megacities were mainly chemical fertilizer and municipal waste. Furthermore, isotopic enrichment was also confirmed in Manila and Jakarta, suggesting a denitrification trend. However, in Bangkok, nitrate concentrations in the groundwater were low because of the anoxic condition, but the obvious isotopic enrichment was not indicated. This suggests that the

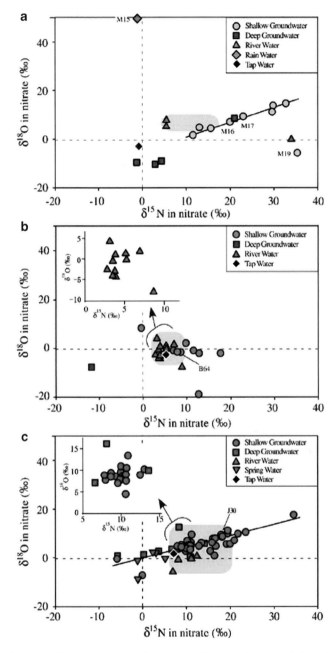

Fig. 9.9 Relationship between oxygen and nitrogen stable isotopic ratios of nitrate in Manila (**a**), Bangkok (**b**), and Jakarta (**c**) (Umezawa et al. 2009a)

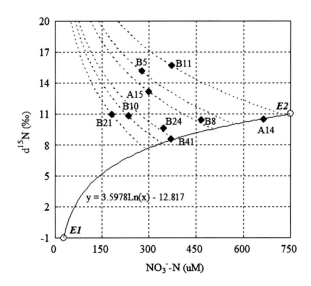

Fig. 9.10 NO$_3^-$-N and δ^{15}N relationships in groundwater of Jakarta. E1 is a spring water in a forest headwater, E2 is nitrate polluted groundwater with high nitrate concentration (Saito et al. 2009)

nitrate load to the groundwater is small. The thick clay layer which covers the aquifers in Bangkok alluvial plain creates a very flat topography. This clay layer would prevent nitrate leaching from the surface to the saturated zone.

Saito et al. (2009) estimated the mixing and denitrification rate in the groundwater of Jakarta, using these data, as shown in Fig. 9.10. The mixing process causes the dilution of nitrate-polluted groundwater (E2) by water with very low nitrate concentration such as spring water from a forest headwater or rain water (E1). The mixing curve (solid line) between E1 and E2, and the isotopic enrichment curves (broken lines) from the denitrification process are shown in Fig. 9.10. They estimated the mixing ratio of the polluted groundwater and enrichment ratio by denitrification to be from 50% to 100%, and from 0% to 50%, respectively.

Hosono et al. (2009) indicated nitrate and sulfate concentrations in groundwater of Seoul in 2005 (Fig. 9.11). They compared it with the results from 1996 and 1997 by Kim (2004). The concentration level was confirmed to decline significantly in the 9 years from 1996 to 2005. However, the nitrate concentrations at some points exceeded the environmental standard. In addition, Nakaya et al. (2009) confirmed the nitrate concentration in groundwater of Osaka. Figure 9.12 shows the spatial distributions in nitrate concentrations of shallow groundwater (G1) and the second aquifer (G2; alluvial and diluvial formation with approximately 30 m depth) in Osaka. The nitrate concentrations were high in the shallow groundwater and in a recharge area. There were some points which exceeded the environmental standard.

Based on these studies, we can confirm the nitrate pollution in shallow groundwater and recharge areas, except in Bangkok. We need to monitor its transport and diffusion in groundwater in the future. However, the denitrification trend was also confirmed. It suggests that nitrate transport to discharge areas is small.

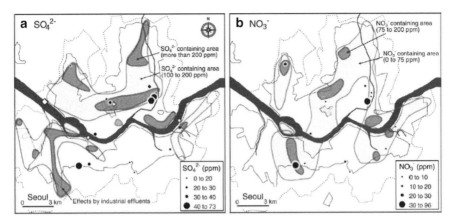

Fig. 9.11 Spatial distribution of sulfate (**a**) and nitrate (**b**) in the groundwater of Seoul (Hosono et al. 2009)

9.5   Trace Metal Pollution

Trace metal pollution had been reported by many researches from various sources such as mines, industrial waste, air pollution, and minerals in aquifers. There are notably many sources in coastal urban areas. For example, the accumulation of trace metals in coastal marine sediments has been confirmed in various areas (Hoshika and Shiozawa 1988; Williams et al. 2000; etc.). These mainly originate from urban air pollution, industrial waste, and surface soil pollutants rather than from direct marine pollution. Table 9.3 shows the variation from the 1890s to the 1990s of trace metal flux into Jakarta Bay, estimated by the analysis of a 40-cm-thick surface sediment core, taken from the sea floor. The peak of copper (Cu) flux was in the 1990s while zinc (Zn) and lead (Pb) peaked in the 1980s. These results suggest that the increase in pollution agrees with the intensive rise in industrial activity as well as population.

Also in Japanese coastal area, Trace metal pollution has also been confirmed in the coastal area of Japan (Hoshika and Shiozawa 1988; Hoshika et al. 1991). Figure 9.13 shows the history of Cu and Zn loads into sediments in various regions of the Seto Inland Sea (Hoshika et al. 1991). There are many industrial or populated areas such as Osaka, Okayama, and Hiroshima around the central area and Yamaguchi and Kita-Kyusyu in the south-west area around the Seto Inland Sea. The values increased from 1800 to 1960, but they have decreased since 1970. These trends are similar to the COD variation in the surface water of Osaka Bay in Fig. 9.3.

In addition, As pollution in groundwater has been confirmed in various megacities (Ito et al. 2003; Hosono et al. 2010; etc.). The pollution is caused by the leaching of the mineral from sediment in aquifers. Figure 9.14 shows the relationship between nitrate and arsenic concentrations in various Asian megacities (Hosono et al. 2010). The high As concentrations were detected in the aquifers with

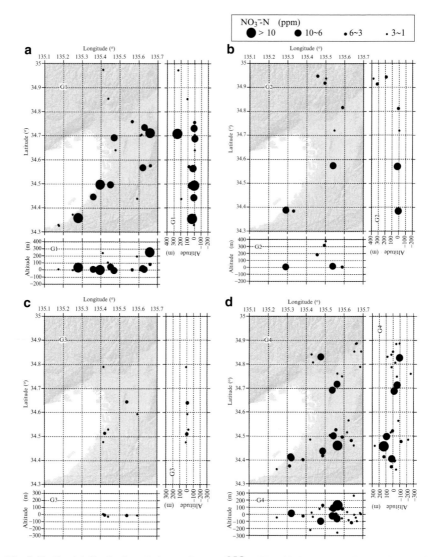

Fig. 9.12 Spatial distribution of nitrate nitrogen (NO_3^--N) with So_4>1ppm of groundwaters for aquifers. (**a**) G1, (**b**) G2, (**c**) G3 and (**d**) G4 (Nakaya et al. 2009)

Table 9.3 Trace metal flux into Jakarta Bay estimated by sediment core analysis. ($\mu g\,cm^{-2}\,year^{-1}$) (Williams et al. 2000)

Depth (cm)	Age (year)	Cu flux	Zn flux	Pb flux
0–1	0–2.6	1.68	3.51	1.12
5–6	16	1.15	4.96	2.59
10–11	29	0.58	4.09	1.95
20–21	56	0.83	1.40	1.94
30–31	82	1.05	1.83	1.57
40–41	109	0.79	0.48	0.24

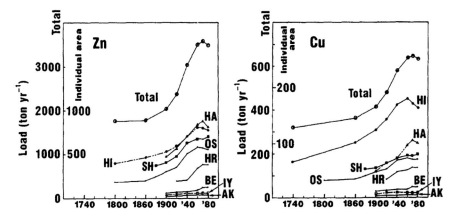

Fig. 9.13 History of Cu and Zn loads into sediments of the various regions of the Seto Inland Sea (Hoshika et al. 1991). *OS* Osaka (east), *HA* Harima (central east), *HI* Hiuchi (central south), *AK* Aki (central north), *HR* Hiroshima (central west), *SH* Suoh (west), *IY* Iyo (south west), *BE* Beppu (south west)

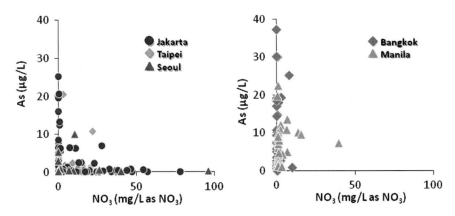

Fig. 9.14 Relationship between nitrate and arsenic concentrations in various Asian megacities (Hosono et al. 2010)

a reductive condition as well as a high ammonium concentration, such as in Bangkok (Umezawa et al. 2009a), but the nitrate concentration indicated an opposite trend. In general, a reductive condition in groundwater is related to the residence time and the rate of oxygen consumption, and depends on the topographic gradient, annual rainfall, and temperature (Drever 1988; Appelo and Postma 2005).

9.6   Chloride Pollution

Chloride originates mainly in the sea and is a necessary component for living organisms. However, the oversupply of salinity to soil and groundwater causes salinization, which is an environmental problem. Groundwater salinization occurs with over-pumping of deep coastal groundwater. The sources of salinity in groundwater are present seawater, and salt in alluvial sediment; the former is apparent as sea water intrusion and the latter is as a supply of fossil salt or palaeo-salt. Groundwater salinization had been confirmed in some coastal megacities (Iwatsu et al. 1960; Onodera et al. 2009; etc.). Normally, water with chloride concentrations higher than 250 mg l^{-1} is unsuitable to drink, according to the environmental standard of the WHO.

Iwatsu et al. (1960) indicated the salinization of coastal shallow groundwater with the highest chloride concentration of 5,000 mg l^{-1} in the Osaka metropolitan area during the period of lowest groundwater potential. The undrinkable water zone expanded from the coastline up to 8-km inland. In this case, sea water intrusion would be the main process of groundwater salinization. In addition, Onodera et al. (2009) showed the salinization of deep groundwater with the highest chloride concentration of 3,500 mg l^{-1} in Jakarta groundwater (Fig. 9.15b). The direction of groundwater flow was downward from the surface to the middle aquifer at a depth of 150 m in the coastal zone (Fig. 9.15a). The concentration was higher in the middle aquifer than in the shallow and deeper aquifers. This suggests seawater intrusion. However, the concentration in the shallow aquifer was still high compared with the environmental standard. In this case, the palaeo-salt leaching from the surface alluvial fine sediment could be suggested as the main process of shallow groundwater salinization in the area covered by the alluvial clay.

Figure 9.16 shows the sources of salt in groundwater in the western part of Nagoya metropolitan area (Yamanaka and Kumagai 2006). This was estimated by using the sulfur isotopic ratio. The present seawater contributed the salt in the area near the river; on the other hand, fossil salt was the contributor in the inter-river area. This suggests the effect of sea water intrusion through the river.

9.7   Complex Problem of Contaminant Transport

The change of groundwater flow affects contaminant transport. Groundwater is naturally recharged in the highlands and is discharged to the lowlands (Tóth 1963; Freeze and Cherry 1978; Domenico and Schwartz 1990). Based on this concept of groundwater flow, mountains and upland areas correspond to the areas of groundwater recharge in a watershed, whereas coastal zones and lowlands are the areas of discharge. In general, the horizontal hydraulic gradient of groundwater is equal or less than the topographic gradient. If the groundwater head drops more than 10 m in

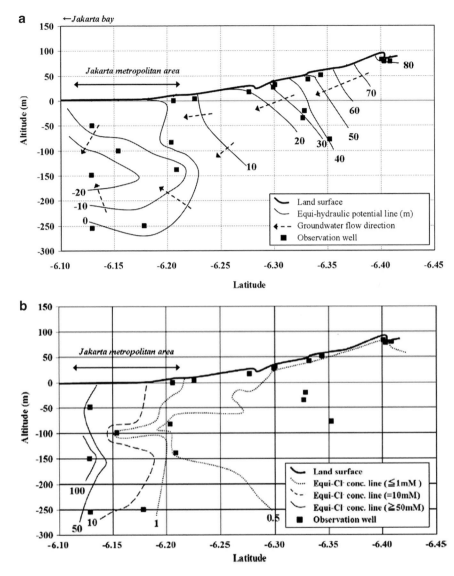

Fig. 9.15 Distribution of flow (**a**) and chloride concentration (**b**) in Jakarta groundwater (Onodera et al. 2009)

flat continental land or coastal alluvial plains, both horizontal and vertical hydraulic gradients change significantly. Kamra et al. (2002) have reported vertical contaminant transport in groundwater after a drop in water level caused by pumping extraction. As stated above, intensive groundwater pumping with urbanization has caused a

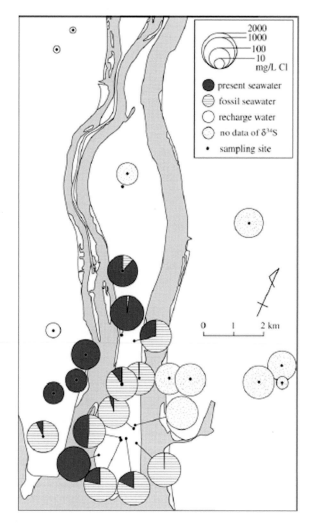

Fig. 9.16 Source of salinization in the western part of Nagoya metropolitan area (Yamanaka and Kumagai 2006)

serious decline in water levels and variations in flow and contaminant transport. For example, the water level decreased more than 30 m from the 1950s to the 1960s in the Tokyo and Osaka aquifers in Japan (Environment Agency Japan 1996). Such decline of groundwater potential causes the change of groundwater flow direction from lateral to downward. As a result, the groundwater will be difficult to discharge to the sea and it would cause the dispersion of contaminants and expansion of the polluted area in the land. Protano et al. (2000) reported the distribution of Hg contamination and its expansion in an area of declining groundwater potential.

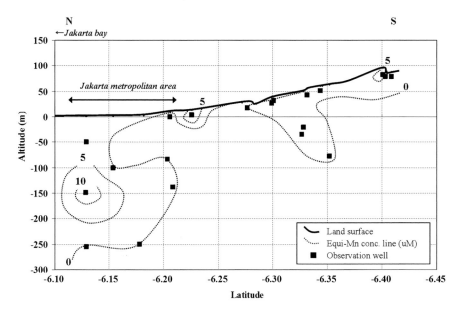

Fig. 9.17   The distribution of Mn concentration in groundwater of Jakarta (Onodera et al. 2009)

The distribution of manganese (Mn) concentration in groundwater on the north–south transect in Jakarta is shown in Fig. 9.17 (Onodera et al. 2009). The concentration was highest around the coastal area, with more than 10 µM at a depth of 150 m, and the hydraulic potential was lowest at this point (Fig. 9.15a). These observations suggest that the high Mn concentration was caused by the intrusion of both shallow groundwater and deep groundwater.

The relation between the difference in water level depth and nitrate-nitrogen (NO_3^--N) concentration in the PD and NL aquifers in Bangkok is shown in Fig. 9.18. PD is the uppermost productive aquifer with a depth of 100 m and NL is the second aquifer at a depth of 150 m underlain by the PD aquifer. Here, the negative value in the horizontal axis means that the groundwater flows downward, and the positive value means the flow direction is upward. The NO_3^--N concentration was lower than the limit of detection (<7.0 µM) at most of the sampling boreholes due to the denitrification, as shown in Fig. 9.9 (Umezawa et al. 2009a). However, high NO_3^--N concentrations of >150 µM were detected at some observation boreholes (No. 61 and No. 78), these are located in the metropolitan area. The certain origin of NO_3^--N is mostly unknown, as the groundwater age was estimated to be more than 10,000 years by Sanford and Buapeng (1996). However, there is the possibility that an inflow of domestic wastewater or mineralization of ammonium-nitrogen (NH_4^+-N) have an influence on it. In Jakarta, NO_3^--N concentrations were also relatively high (>200 µM) in the shallow groundwater (<–50 m) of the urban area. As stated above, the concentrations of Mn and NO_3^--N were relatively high in the coastal urban area in both Bangkok and Jakarta, and these areas were also

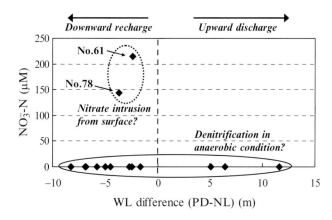

Fig. 9.18 Relationship between water level difference of PD and NL aquifers and nitrate concentrations in PD and NL aquifers (Onodera et al. 2009). The PD aquifer is at a depth of around 100 m, the NL aquifer is around 150 m deep

characterized by downward groundwater flow. These observations suggest the intrusion of these contaminants from shallow groundwater to the deep groundwater and imply the accumulation of contaminants in deeper aquifers.

9.8 Effect on Marine Environment

As indicated in Fig. 9.3, the sea surface water quality in Osaka Bay had not recovered after 1970, compared with the river water quality. This suggests the delayed discharge effects of pollutants from the land into the sea as well as the storage effect of the sea water body and bottom sediments in the bay. Based on the properties of pollutant accumulation in groundwater described above, it is important to recognize the possibility of contaminant transport with the discharge of deep groundwater into the sea after the recovery of its potential in coastal areas. In the Osaka plain, most of the groundwater potential has already recovered in and around the central area. In addition, Ishitobi et al. (2007) showed the submarine groundwater discharge (SGD) was similar to the river discharge in the west coast of the Osaka metropolitan city, using the hydrologic observation as confirmed at other areas by previous researches (Taniguchi et al. 2002; Burnett et al. 2006; etc.).

On the other hand, in the Seto Inland Sea, a huge amount of contaminant was accumulated in the seabed before 1970 from surface water pollution as shown in Fig. 9.13. In addition, many researches (Burnett et al. 2006; Onodera et al. 2007; Taniguchi et al. 2007; Umezawa et al. 2007) have indicated the leaching of nutrients through the recirculation of sea water along the nearshore area. Figure 9.19 shows the relationship between Cl^- and NO_3^- concentration in sea water, terrestrial groundwater and porewater below the tidal slope along the coastal area of the Seto Inland Sea (Onodera et al. 2007). In this figure, most of the porewater samples were plotted away

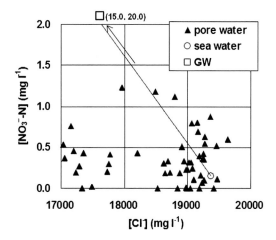

Fig. 9.19 Relationship between Cl⁻ and NO₃⁻-N concentrations in sea water, terrestrial groundwater and porewater below tidal slope (Onodera et al. 2007)

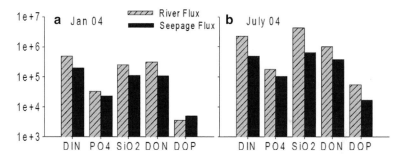

Fig. 9.20 Material transport by groundwater and river water during dry season (**a**) and wet season (**b**) (Burnett et al. 2007)

from the line that runs between the groundwater and seawater. This result indicates both the attenuation and leaching process of nitrate. The attenuation process occurred from the terrestrial area to the tidal slope, but the leaching process occurred in offshore areas. The leaching rate was accelerated by the recirculation of seawater.

In case of Thailand Bay bordering Bangkok, the total nutrient discharge by seepage in the coastal zone including the seawater recirculation was estimated to be similar to the river flux by Burnett et al. (2007) as shown in Fig. 9.20.

It is important to confirm the spatial variation of submarine groundwater discharge for the evaluation of the delayed contaminant transport by discharge of terrestrial groundwater and leaching from the sediment. Figure 9.21 shows the spatial variation of radon (²²²Rn) concentrations in Jakarta Bay (Umezawa et al. 2009b). It is generally reflected in the variation of groundwater discharge, according to Burnett et al. (2006, 2009). Based on the result, the groundwater discharge is

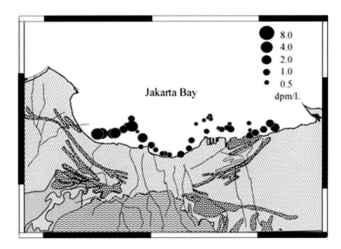

Fig. 9.21 Distribution of ^{222}Rn concentrations along the coastal line in Jakarta Bay (Umezawa et al. 2009a, b)

higher in the west coast compared with that in other areas. Part of the western area is covered by natural vegetation with a relatively large topographic gradient.

The topographic gradient is a good indicator of groundwater discharge. Shimizu et al. (2009) estimated the spatial variation of groundwater discharge in an area of the Seto Inland Sea at intervals of 50 m, using a topographic model. They confirmed a good correlation between the topographic gradient of the coastal area and groundwater discharge.

9.9 Conclusions

In this chapter, the various subsurface pollutants associated with urbanization were confirmed, such as nitrate, trace metals, and chloride. In general, the vulnerability of megacities to pollution is found to be high but the variation for each city is not clear. In this chapter, an indication of this is suggested by the intensities of surface pollution (flow) and subsurface pollution (accumulation). These intensities are controlled by human impact as well as the natural background, as shown in Fig. 9.22. For example, the emission and load of pollutants increase with the population creating surface pollution in the first stage of city development. Then, subsurface pollution occurs as surface pollutants are transported to the groundwater. In addition, groundwater abstraction affects the intrusion of surface pollution to deep groundwater. On the other hand, contamination and attenuation processes related to groundwater flow conditions are controlled by the natural features such as topography, geology, watershed area, and natural recharge or climate. In the case of Bangkok, the topographic gradient is very small and the natural recharge is also small. Hence, the attenuation rate of nitrate by denitrification is relatively large, while chloride contamination by palaeo-salt is serious.

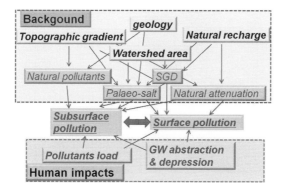

Fig. 9.22 Controlling factors of vulnerability to pollution

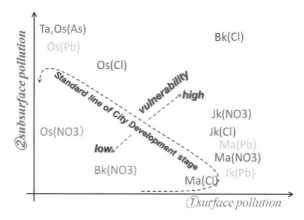

Fig. 9.23 Vulnerability of coastal megacities to different pollutants. *Bk* Bangkok, *Jk* Jakarta, *Ma* Manila, *Os* Osaka, *Ta* Taipei

The estimated vulnerability of Asian megacities to the various pollutants is shown in Fig. 9.23. The horizontal axis represents the surface pollution and the vertical axis represents the subsurface pollution. It is suggested that the vulnerability to each pollutant is determined by the natural environment and the developing stage of each city. These components control the groundwater flow conditions related to the oxidation-reduction reaction and the pollutant load by human activity. In this diagram, the vulnerability is higher in the top right area and is low at the bottom left area. A city in the first stage of development, going through an intensive growth of population, is plotted at the bottom right area, whereas the cities in the third stage, which are in the developed stage with infrastructure, are plotted on the top left area (Fig. 9.7). The estimated vulnerability of Jakarta to nitrate, trace metals, and chloride contamination is high, Bangkok is extremely vulnerable to chloride but less vulnerable to nitrate, and Osaka is highly vulnerable to As and chloride.

Fig. 9.24 Large amounts of garbage and trash in river (**a**) washed onto beach (**b**) in Jakarta

In addition, water pollution changes from the first stage to the third stage of city development as shown in Fig. 9.7. The garbage and trash problems have not been discussed in this chapter. Garbage and trash are transported downstream by rivers from upstream sources (Fig. 9.24a), all the way to the sea and beaches (Fig. 9.24b). Before 1970, in the first stage of development of Tokyo and Osaka, garbage did not include plastic; however plastic constitutes the main part of garbage in Jakarta and Manila at present. These problems have to be studied in detail.

Acknowledgments I thank the staffs and students of the RIHN project and students of RM project in Hiroshima University. Especially, I would like to give the special thanks to the main material group members, Dr. T. Hosono, Dr. Y. Umezawa, Prof. T. Nakano, Prof. S. Nakaya, Dr. M. Saito, and Mr. Y. Shimizu, and Project leader, Prof. M. Taniguchi, and foreign counterpart members.

References

Abidin HZ, Andreas H, Djaja R, Darmawan D, Gamal M (2007) Land subsidence characteristics of Jakarta between 1997 and 2005, as estimated using GPS surveys. GPS Solut. doi:10.1007/s10291-007-0061-0

Appelo CAJ, Postma D (2005) Geochemistry, groundwater and pollution. AA Balkema Publishers, Leiden

Böhlke JK, Denver JM (1995) Combined use of groundwater dating, chemical, and isotopic analyses to resolve the history and fate of nitrate contamination in two agricultural watersheds, Atlantic coastal plain, Maryland. Water Resour Res 31:2319–2340

Böhlke JK, Verstraeten IM, Kraemer TF (2007) Effects of surface-water irrigation on sources, fluxes, and residence times of water, nitrate, and uranium in an alluvial aquifer. Appl Geochem 22:152–174

Brinkhoff T (2010) City population. http://www.citypopulation.de/. Accessd on 01/01/2010

Burnett WC, Aggarwal PK, Aureli A, Bokuniwicz H, Cable JE, Charette MA, Kontar E, Krupa S, Kulkarni KM, Loveless A, Moore WS, Oberdorfer JA, Oliveira J, Ozyurt IN, Povinec P, Privitera AMG, Rajar R, Ramessur RT, Schollten J, Stieglitz T, Taniguchi M, Turner JV (2006) Quantifying submarine groundwater discharge in the coastal zone via multiple methods. Sci Total Environ 367:498–543

Burnett WC, Wattayakorn G, Taniguchi M, Dulaiova H, Sojisuporn P, Rungsupa S, Ishitobi T (2007) Groundwater derived nutrient inputs to the Upper Gulf of Thailand. Cont Shelf Res 27:176–190

Burnett WC, Chanyotha S, Wattayakorn G, Taniguchi M, Umezawa Y, Ishitobi T (2009) Underground sources of nutrient contamination to surface waters in Bangkok, Thailand. Sci Total Environ 407:3198–3207

Burt TP, Heathwaite AL, Trudgill ST (1993) Nitrate: processes, patterns and management. Wiley, New York

Domenico PA, Schwartz FW (1990) Physical and chemical hydrogeology. Wiley, New York

Drever JI (1988) The geochemistry of natural waters, 2nd edn. Prentice Hall, Englewood Cliffs

Environment Agency Japan (1969) White paper: quality of the environment in Japan. Printing Bureau, Ministry of Finance Japan, Tokyo

Environment Agency Japan (1996) White paper: quality of the environment in Japan. Printing Bureau, Ministry of Finance Japan, Tokyo

Environment Council, Osaka Prefectural Government (2004) Document for Planning Group of Water Quality Monitoring, No. 3. http://www.epcc.pref.osaka.jp/kannosomu/kankyo_singikai/water/giji/index.html

Foster SSD (2001) The interdependence of groundwater and urbanization in rapidly developing cities. Urban Water 3:185–192

Freeze RA, Cherry JA (1978) Groundwater. Prentice Hall Inc., Englewood Cliffs

Gandy CJ, Smith JWN, Jarvis AP (2007) Attenuation of mining-derived pollutants in the hyporheic zone. Sci Total Environ 373:435–446

Hinkle SR, Duff JH, Triska FJ, Laenen A, Gates EB, Bencala KE, Wentz DA, Silva SR (2001) Linking hyporheic flow and nitrogen cycling near the Willamette River: a large river in Oregon, USA. J Hydrol 244:157–180

Hoshika A, Shiozawa T (1988) Mass balance of heavy metals in the Seto Inland Sea, Japan. Mar Chem 24:327–335

Hoshika A, Shiozawa T, Kawana K, Tanimoto T (1991) Heavy metal pollution in sediment from the Seto Inland Sea, Japan. Mar Pollut Bull 23:101–105

Hosono T, Ikawa R, Shimada J, Nakano T, Saito M, Onodera S, Lee K, Taniguchi M (2009) Human impacts on groundwater flow and contamination deduced by multiple isotopes in Seoul City, South Korea. Sci Total Environ 407:3189–3197

Hosono T, Nakano T, Shimizu Y, Onodera S, Taniguchi M (2010) Hydrogeological constraint on nitrate and arsenic contamination in Asian metropolitan groundwater. Hydrological Processes (in press)

Howard KWF (1985) Denitrification in a major limestone aquifer. J Hydrol 76:265–280

Ishitobi T, Taniguchi M, Umezawa Y, Kasahara S, Onodera S, Miyaoka K, Hayashi M, Hayashi M (2007) Investigation of submarine groundwater discharge using several methods in the inter-tidal zone. IAHS Publ 312:60–67

Ito H, Masuda H, Kusakabe M (2003) Variations of arsenic contents in groundwater and its factors in North Settsu region, Osaka, Japan. J Groundwater Hydrol 45:3–18 (Japanese with English abstract)

Iwatsu J, Tsurumaki D, Ichihara Y (1960) Qualities and some issues of groundwater in a west part of Osaka City, Japan. J Groundwater Hydrol 2:1–14 (Japanese with English abstract)

Jiang Y, Kirkman H, Hua A (2001) Megacity development: managing impacts on marine environments. Ocean Coast Manag 44:293–318

Kamra SK, Lal K, Singu OP, Boonstra J (2002) Effect of pumping on temporal changes in groundwater quality. Agric Water Manage 56:169–178

Kendall C, McDonnell JJ (1998) Isotope tracers in catchment hydrology. Elsevier Science, Amsterdam

Kim YY (2004) Analysis of hydrochemical processes controlling the urban groundwater system in Seoul area, Korea. Geosci J 6:319–330

Nakatsuji K (1998) Water environment in coastal area. In: Takahashi Y, Kawada K (eds) Water cycle and catchment environment. Iwanami Shoten, Tokyo, pp 83–107 (in Japanese)

Nakaya S, Mitamura S, Masuda H, Uesugi K, Hondate Y, Kusakabe M, Iida T, Muraoka H (2009) Recharge sources and flow properties of groundwater in Osaka Basin estimated by using environmental tracers and water quality, Japan. J Groundwater Hydrol 51:15–41 (Japanese with English abstract)

Onodera S, Saito M, Hayashi M, Sawano M (2007) Nutrient dynamics with interaction of groundwater and seawater in a beach slope of steep island, western Japan. IAHS Publ 312:150–158

Onodera S, Saito M, Sawano M, Hosono T, Taniguchi M, Shimada J, Umezawa Y, Lubis RF, Buapeng S, Delinom R (2009) Effects of intensive urbanization on the intrusion of shallow groundwater into deep groundwater. Examples from Bangkok and Jakarta. Sci Total Environ 407:3209–3217

Postma D, Boesen C, Kristiansen H, Larsen F (1991) Nitrate reduction in an unconfined sandy aquifer: water chemistry, reduction processes, and geochemical modeling. Water Resour Res 27:2027–2045

Protano G, Riccobono F, Sabatini G (2000) Does salt water intrusion constitute a mercury contamination risk for coastal fresh water aquifers? Environ Pollut 110:451–458

Saito M, Onodera S, Umezawa Y, Hosono T, Shimizu Y, Delinom R, Taniguchi M (2009) Evaluation of nitrate attenuation potential in the groundwater of Jakarta metropolitan area, Indonesia. IAHS Publ 329:305–310

Sanford WE, Buapeng S (1996) Assessment of a groundwater flow model of the Bangkok Basin, Thailand, using carbon-14-based ages and paleohydrology. Hydrogeol J 4:26–40

Shimizu Y, Onodera S, Saito M (2009) Estimation of spatial distribution in submarine groundwater discharge, using 50 m DEM and GIS model: an example applied in the central area of Seto Inland Sea, Japan. J Limnol 70:129–139 (Japanese with English abstract)

Shindo J, Okamoto K, Kawashima H (2003) A model-based estimation of nitrogen flow in the food production-supply system and its environmental effects in East Asia. Ecol Modell 169:197–212

Slomp CP, Cappellen PV (2004) Nutrient inputs to the coastal ocean through submarine groundwater discharge: controls and potential impact. J Hydrol 295:64–86

Taniguchi M, Burnett WC, Cable JE, Turner JV (2002) Investigations of submarine groundwater discharge. Hydrol Process 16:2115–2129

Taniguchi M, Burnett WC, Dulaiova H, Siringan F, Foronda J, Wattayakorn G, Rungsupa S, Kontar EA, Ishitobi T (2007) Groundwater discharge as an important land-sea pathway into Manila Bay, Philippines. J Coast Res 24(1A):15–24

Tesoriero AJ, Liebscher H, Cox SE (2000) Mechanism and rate of denitrification in an agricultural watershed: electron and mass balance along groundwater flow path. Water Resour Res 36:1545–1559

Tóth J (1963) A theoretical analysis of groundwater flow in small drainage basins. J Geophys Res 68:4795–4812

Tsunekawa A (1998) Comparison of world urban environment, using environmental index. In: Takeuchi K, Hayashi Y (eds) Global environment and mega-cities. Iwanami Shoten, Tokyo, pp 29–56 (in Japanese)

Umezawa Y, Ishitobi T, Rungsupa S, Onodera S, Yamanaka T, Yosimizu C, Tayasu I, Nagata T, Wattayakorn G, Taniguchi M (2007) Evaluation of fresh groundwater contributions to the nutrient dynamics at shallow subtidal areas adjacent to metro-Bangkok. IAHS Publ 312:169–179

Umezawa Y, Hosono T, Onodera S, Siringan F, Buapeng S, Delinom R, Yoshimizu C, Tayasu I, Nagata T, Taniguchi M (2009a) Sources of nitrate and ammonium contamination in groundwater under developing Asian megacities. Sci Total Environ 407:3219–3231

Umezawa Y, Onodera S, Ishitobi T, Hosono T, Delinom R, Burnett W, Taniguchi M (2009b) Effect of urbanization on the groundwater discharge into Jakarta Bay. IAHS Publ 329:233–240

United Nations (1999) World urbanization prospects. United Nation, New York p.160, 52l

Williams TM, Rees JG, Setiapermana D (2000) Metals and trace organic compounds in sediments and waters of Jakarta Bay and the Pulau Seribu Complex, Indonesia. Mar Pollut Bull 40:277–285

World Bank (1997) World development indicators, CD-ROM. World Bank, Washington

Yamanaka M, Kumagai Y (2006) Sulfur isotope constraint on the provenance of salinity in a confined aquifer system of the southwestern Nobi Plain, central Japan. J Hydrol 325:35–55

Part IV
Subsurface Thermal Anomalies Due to Global Warming and Urbanization

Chapter 10
Comparisons Between Air and Subsurface Temperatures in Taiwan for the Past Century: A Global Warming Perspective

Chieh-Hung Chen, Chung-Ho Wang, Deng-Lung Chen, Yang-Yi Sun, Jann-Yenq Liu, Ta-Kang Yeh, Horng-Yuan Yen, and Shu-Hao Chang

Abstract Air and sea surface temperature increases due to global warming have been widely observed around the world at various rates. This temperature rising has also been documented in many subsurface records recently. The air-ground temperature coupling system introduces an important factor in disturbing the original thermal balance and provides a new dimension to comprehend the effects of global warming on the Earth system. Ten meteorological stations of Central Weather Bureau in Taiwan that have been routinely measured for air (1.5 m above the ground) and subsurface (at depths of 0, 5, 10, 20, 30, 50, 100, 200, 300 and 500 cm below the ground) temperatures are used for in-depth comparison in this study. These stations have a mean observation period of 82 years (as of 2008) to provide good coverage for a preliminary examination of air-ground temperature coupling relationship in a century

C.-H. Chen (✉), C.-H. Wang, and S.-H. Chang
Institute of Earth Sciences, Academia Sinica, 128, Sec. 2, Academia Rd.,
Nangang, Taipei 11529, Taiwan, ROC
e-mail: nononochchen@gmail.com

D.-L. Chen
Penghu Station, Central Weather Bureau, 2, Xinxing Rd., Magong, Penghu 88042, Taiwan, ROC

Y.-Y. Sun
Institute of Space Science, National Central University, 300, Jhongda Rd.,
Jhongli, Taoyuan 32001, Taiwan, ROC

J.-Y. Liu
Institute of Space Science, National Central University, 300, Jhongda Rd.,
Jhongli, Taoyuan 32001, Taiwan, ROC
and
Center for Space and Remote Sensing Research, National Central University,
300, Jhongda Rd., Jhongli, Taoyuan 32001, Taiwan, ROC

T.-K. Yeh
Institute of Geomatics and Disaster Prevention Technology, Ching Yun University,
229, Jianxing Rd., Jhongli, Taoyuan 32097, Taiwan, ROC

H.-Y. Yen
Institute of Geophysics, National Central University, 300, Jhongda Rd.,
Jhongli, Taoyuan 32001, Taiwan, ROC

M. Taniguchi (ed.), *Groundwater and Subsurface Environments: Human Impacts in Asian Coastal Cities*, DOI 10.1007/978-4-431-53904-9_10, © Springer 2011

scale. Results show that patterns and variations of air and subsurface temperature are quite different among stations in Taiwan. In general, air and subsurface temperatures exhibit consistent linear trends after 1980 due to accelerating global warming, but display complex and inconsistent tendencies before 1980. When surface air temperature is subtracted from subsurface one, the differences in the eastern Taiwan are generally larger than those in the western Taiwan. This observation is possibly caused by (1) heat absorption of dense high-rise buildings, and/or (2) cut off heat propagating into deep depths in the urban area of western Taiwan. By comparing temperature peaks at various layers from shallow to deep, rates of thermal propagation can be estimated. The distinct time shifts among stations suggest that thermal propagations have to be taken into account when constructing historical temperature records.

10.1   Introduction

Concentration increase of greenhouse gases in the atmosphere is generally regarded as the main cause of the global warming (IPCC 2007). Productions and surface emissions of greenhouse gases through biogeochemical processes in soil are closely modulated by air temperature (Lloyd and Taylor 1994; Risk et al. 2002a, b; Luo et al. 2001). Therefore, study of the air-ground temperature coupling is very important for climate change research (Beltrami and Kellman 2003). Many studies indicated that historical changes for ground surface temperature (GST; i.e., relatively shallow subsurface temperature) over a large range of spatial and temporal scales can be reconstructed by current measurements for temperature-depth profiles (Pollack and Huang 2000; Huang et al. 2000; Beltrami 2001a). The reconstructions are generally used to compare with historical surface air temperature (SAT) and good agreements have given credence to the relationship between SAT and GST (Huang et al. 2000; Harris and Chapman 2001; Beltrami 2002). However, the reconstructions, which have been comprised by underground temperatures at deep depths, present smoothed versions for GST history, signals such as diurnal and/or seasonal cycles would be generally lost due to a lack of short temporal variations. To evaluate causes of short-term changes in subsurface temperature, climatic variations, such as snow cover, soil freezing, evapotranspiration, and vegetations, can be well linked to GST with a good correlation if data sets are available (Baker and Ruschy 1993; Putnam and Chapman 1996; Beltrami 2001b; Zhang et al. 2001; Baker and Baker 2002; Smerdon et al. 2003, 2006).

In short, variations of subsurface temperature are mainly comprised by responses of solar radiation from the space to Earth, thermal conduction from inner Earth outward to the surface, and climatic effects near the Earth surface. Because of a long propagation distance that passed strata as a low-pass filter, the inward radiation and outward thermal conduction dominate the long-term changes of subsurface temperature. Over last decades, global warming has significantly shown on the rising trend of SAT on earth surface (IPCC 2007). This additional element enforces significant changes on the complex air-ground temperature coupling system, and sheds more lights on the relationship between GST and SAT.

Because the influence of environmental changes may sustain for a long temporal period, data recorded covering a few decades are often insufficient for long-term analyses. In this study, ten meteorological stations in Taiwan (Table 10.1) with an average observation period of 82.2 years for GST records are selected. For simplification, we compare linear trends of SAT with those of GST to study effects caused by the inward source. Meanwhile, we compute the annual changes between GST and SAT to comprehend the characteristics at each site, and further evaluate the propagation time for thermal conduction from shallow to deep depth in Taiwan.

10.2   Data

For the observation practice, sporadic measurements of GST and SAT in Taiwan could be traced to the late Ching Dynasty of China. In 1896, Taiwan Island was transferred to Japan; meteorological stations were gradually established one by one under Japanese governance and officially started the systematic observation. After 1945, Taiwan was back to the jurisdiction of Republic of China, the Central Weather Bureau (CWB) took over existing stations and expanded more for continuous observation. Till to 2008, observed periods of ten selected stations range from 41 years in Chiayi to 109 years in Hengchun. Hence, temperature records with such a long observation period are unique and valuable. In addition, these stations are distributed rather evenly in Taiwan. Taking the Taipei station as a reference, GST observation was initiated in this site of northern Taiwan from 1930 (Fig. 10.1 and Table 10.1). From the north to south, stations Hsinchu, Taichung, Chiayi and Tainan are positioned in western side, and stations Ilan, Hualien and Taitung are situated in the eastern Taiwan, respectively (Fig. 10.1). At the southern tip of Taiwan locates the Hengchun station (starting in 1900) with a typical tropics climate.

On the other hand, intense interaction between the Philippine Sea plate and the Eurasian plate results the complex topography in Taiwan and raises the Central Range up to an elevation of approximately 4,000 m (Ho 1988). Station Jiyuehtan was built in the central Taiwan with the altitude of about 800 m and experiences much less influence of human activity. In terms of geology, these stations are all set on the Quaternary sediment resulting from intensity erosion of Tertiary strata in high altitudes (Ho 1988). Because the observation depth is confined to a depth within 5 m from surface, geological background is very similar for all stations. In short, temperature changes under the distinct climate regions (subtropics vs. tropics), different altitudes, urban and rural environments can be evaluated by comparing records in these stations.

GST and SAT records generate a good contrast within one site between these two observation data sets. SAT is recorded with a high sampling interval of 1 min at a height of 1.5 m above the surface. On the other hand, GST data are obtained once a day at depths of 0, 5, 10, 20, 30, 50, 100, 200, 300 and 500 cm (Table 10.1). For a long observation, data continuity is very important and has to be taken into account. In this work, ten observation depths in the ten stations would yield 100 GST data series, 61 of them have the percentages larger than 99%, especially at

Table 10.1 Location, observation periods (year) and data continuity (%) in the stations. Note that the year shown in Table 10.1 is a comprehensive result from entire subsurface observation depths. The percentages of the data continuity are computed using the observation periods at each depth

Stations	Long.	Lat.	Periods	0 cm	5 cm	10 cm	20 cm	30 cm	50 cm	100 cm	200 cm	300 cm	500 cm
Taipei	121.51	25.04	1930–1996	99.4	99.6	99.4	99.4	72.6	99.6	99.6	99.5	97.2	91.0
Hsinchu	121.01	24.83	1939–	100.0	100.0	99.9	100.0	100.0	100.0	100.0	100.0	99.7	99.6
Taichung	120.68	24.15	1901–	99.5	99.9	93.9	89.4	85.9	99.9	99.9	100.0	99.9	88.4
Jiyuehtan	120.9	23.88	1950–	98.8	98.8	98.8	92.2	98.3	98.9	98.9	98.9	98.0	98.0
Chiayi	120.42	23.5	1968–	100.0	100.0	100.0	100.0	100.0	100.0	100.0	100.0	100.0	99.9
Tainan	120.2	23	1900–2001	99.9	99.8	95.3	89.7	88.2	96.0	96.0	100.0	96.0	92.7
Hengchun	120.74	22.01	1900–	98.6	98.3	96.8	74.4	72.6	99.4	99.4	99.2	99.4	99.2
Taitung	121.15	22.75	1902–	100.0	100.0	100.0	100.0	100.0	99.9	98.0	100.0	96.4	97.2
Hualien	121.61	23.98	1922–	100.0	99.9	99.9	99.9	99.8	99.8	99.2	99.9	99.9	100.0
Ilan	121.75	24.77	1936–	99.3	99.9	100.0	98.8	97.8	83.1	83.1	82.6	82.6	83.1

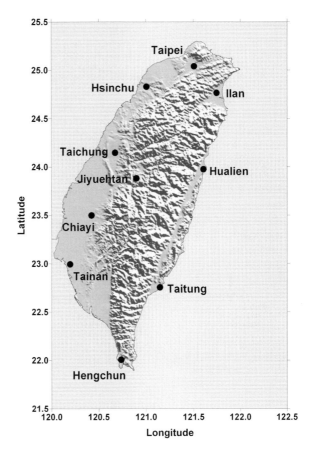

Fig. 10.1 Locations of subsurface temperature stations denoted by *circles*

stations of Hsinchu, Chiayi and Hualien (Table 10.1). In other words, mean data gaps are less than 3 days in 1 year at most stations. In contrast, only 13 data series are smaller than 90% and mainly found in the Ilan station (Table 10.1).

For a better comparison, both minute measurements for SAT and daily records for GST are integrated into mean-annual temperatures. Station Hualien is taken as an example to describe temperature changes in the long temporal domain. Figure 10.2 shows linear trends of the mean GST and SAT at the Hualien station using the least square method (Rao et al. 1999; Wolberg 2005) to examine their relationships. The mean-annual SAT at Hualien ranged between 21.5°C and 24°C from 1922 to 2008, and shows a clear upward trend of 0.14°C/10 year (Fig. 10.2a). On contrast, the mean-annual GST at depth 0 cm displays scattering distribution and reveals an unclear tendency during the entire 87 years (Fig. 10.2b). For the depth of 5 cm, the temperature variations fitted with a linear trend are apparently inappropriate because two discrete patterns are clearly found before and after 1980 (Fig. 10.2c). This suggests a turning point may be in existence approximately in 1980. Regarding the subsurface temperature at depth of 10 cm, data are consistent to a fitting trend of 0.17°C/10 year. In terms of the deeper depths, patterns of the

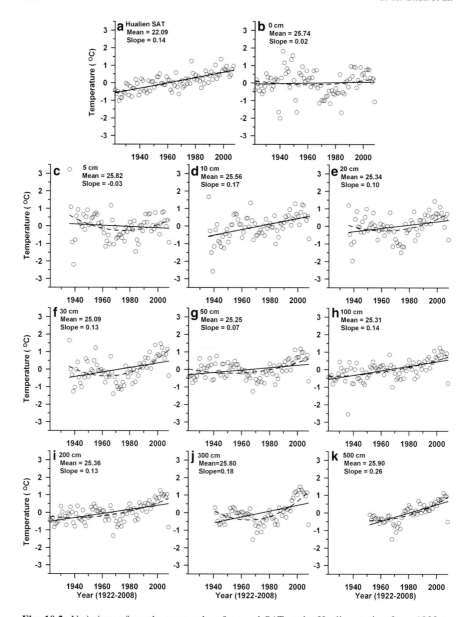

Fig. 10.2 Variations of yearly mean subsurface and SAT at the Hualien station from 1922 to 2008. (**a**) SAT variations; (**b**)–(**k**) subsurface temperatures at depths of 0, 5, 10, 20, 30, 50, 100, 200, 300 and 500 cm, respectively. The *open circles* show the yearly mean temperature and are expressed as departures from long-term means. *Solid lines* and *dashed curves* denote tendencies of yearly mean temperature using the linear trends and second-order functions, respectively. Note that the slop (°C/10 year) indicates the long-term trend of temperature changes within study periods

subsurface temperatures at depths of 20, 30, 50, 100, 200, 300 and 500 cm are roughly similar with that at 5 cm depth, showing that distinctively negative and positive slopes dominate the temperature changes before and after 1980 respectively (Fig. 10.2e–k). In short, the negative tendencies are generally observed before 1980 at deep depths and positive trends found after 1980 at most depths.

Because the year 1980 seems to be a key turning point with respect to environmental change, linear trends of GST and SAT records before and after 1980 are computed to further understand temperature transformations that may be possibly induced by the effect of global warming (Table 10.2). Table 10.2 exhibits statistical results for the linear trends before and after 1980 for ten stations. The increase or decrease trends are regarded as significant when the slopes $\geq 0.1\,°C/10$ year or $\leq -0.1\,°C/10$ year, respectively. SATs at nine stations have conspicuous increasing tendencies after 1980. It is worth to mention that the decrease tendencies are found during the entire observation period only at the Jiyuehtan station. Regarding the subsurface temperature before 1980, the prominent increase trends are only observed at stations with a proportion between 0% and 30%. In contrast, 30% to 63% stations show evident increase trends after 1980. Meanwhile, tendencies before 1980 were subtracted from those after 1980 to study temperature trend shifts. Similarly, difference of the slopes being larger (or smaller) than $0.1\,°C/10$ year (or $-0.1\,°C/10$ year) is determined as significant changes. For SAT, the prominent positive changes are found in most stations, except for stations Jiyuehtan and Chiayi. In a comprehensive survey, the GST and SAT do present strong positive transformations possibly due to the effect of global warming, but these features show inconsistencies in trends for stations of Taipei and Hengchun.

To examine this inconsistent relationship, monthly distributions of long-term subsurface temperatures for various depths (0–500 cm) (MSTs) and for SAT (MAT) are computed respectively. Data of each month are averages of the entire observation period. Here, the Taipei station is taken as an example to understand the relationship between MST and MAT (Fig. 10.3). Figure 10.3a illustrates that MAT is distributed between 15°C and 29°C in this metropolitan station. With the increase of depths, the distribution ranges of MST are sharply reduced down to between 22°C and 24°C at the depth of 500 cm. It is interesting to find that the phases of MST are also correlated with the observation depth. The highest and lowest temperatures observed at shallow levels are gradually departed from July and January through depths, respectively. The time shift of the phase changes between the surface and the depth of 500 cm is about 4 months in Taipei. Moreover, Fig. 10.3b shows the difference given by subtracting MST from MAT at each depth for the Taipei station. The similar patterns between MAT and MST at the shallow depths (≤ 10 cm) yield a small residual value with an average about 0.5°C and are varied within a small range of ± 0.5°C (Fig. 10.3a). The obtained ranges increase with the observation depth of MST and reaches to ± 7.5°C for the deepest measurement (500 cm), which is one order magnitude larger than that of the shallow depth (± 0.5°C).

Figures 10.4 and 10.5 present the averages and ranges of the differences (MST-MAT) at the shallow (≤ 10 cm) and deep (≥ 100 cm) depths respectively, corresponding

Table 10.2 The linear trends (°C/10 years) before and after 1980 and their statistic results at each depth in the stations. Here, D% and I% denote proportions of the stations with decrease and increase transformations of the trends before and after 1980, respectively. Note that T% indicates the percentages of the stations with an increase trends >0.1°C/10 years of subsurface and SAT during the associated period

Stations	SAT		0 cm		5 cm		10 cm		20 cm		30 cm		50 cm		100 cm		200 cm		300 cm		500 cm		Each station	
	1900–1980	1980–2008	1900–1980	1980–2008	1900–1980	1980–2008	1900–1980	1980–2008	1900–1980	1980–2008	1900–1980	1980–2008	1900–1980	1980–2008	1900–1980	1980–2008	1900–1980	1980–2008	1900–1980	1980–2008	1900–1980	1980–2008	D%	I%
Taipei	0.115	0.402	-0.269	-0.109	-0.138	-0.475	-0.247	-0.252	-0.262	-0.437	-0.263	-0.418	-0.187	-0.707	-0.153	-0.581	-0.092	-0.733	-0.101	-0.217	-0.112	-0.090	80%	0%
Hsinchu	0.071	0.289	-0.562	0.051	-0.261	0.014	-0.322	0.049	-0.334	0.049	-0.177	-0.018	0.168	0.285	0.040	0.230	0.049	0.051	0.151	0.189	0.188	0.269	0%	70%
Taichung	0.119	0.438	-0.055	0.139	-0.227	-0.161	-0.195	-0.022	-0.125	-0.030	-0.062	-0.211	0.073	-0.064	0.049	0.026	0.064	0.198	0.047	0.250	0.254	0.308	20%	50%
Jiyuehtan	-0.067	-0.059	-0.276	-0.216	-0.541	-0.103	-0.262	-0.218	-0.104	-0.313	0.067	-0.289	-0.189	-0.137	-0.083	-0.093	-0.103	-0.091					27%	27%
Chiayi	0.697	0.352	0.367	0.116	0.445	0.031	0.548	0.031	0.453	0.090	0.033	-0.025	0.279	0.053	0.172	0.144	-0.268	0.261	-0.444	0.222	-0.446	0.128	50%	30%
Tainan	0.172	0.382	-0.147	0.087	0.082	0.019	0.026	0.019	0.288	0.215	0.087	0.272											0%	30%
Hengchun	0.132	0.232	0.066	-0.371	0.080	-0.129	0.144	-0.130	0.011	-0.276	-0.030	-0.278	0.090	-0.115	0.137	0.034	0.066	0.065	0.050	0.043	0.153	0.092	70%	0
Taitung	0.119	0.247	-0.301	0.236	0.256	0.196	-0.142	0.160	0.001	0.045	-0.047	0.141	-0.007	0.067	-0.051	-0.244	-0.099	-0.032	0.007	-0.082	0.080	-0.043	10%	30%
Hualien	0.135	0.265	-0.087	0.293	-0.224	0.167	0.164	0.176	-0.150	0.281	-0.130	0.470	-0.040	0.343	0.087	0.267	0.028	0.293	-0.104	0.564	0.054	0.217	0%	100%
Ilan	0.082	0.323	-0.271	0.141	0.049	0.530	0.072	0.373	0.067	0.405	0.194	0.368	0.004	0.434	0.007	0.399	0.054	0.256	0.032	0.542	-0.047	0.466	0%	100%
T%	70%	90%	10%	50%	20%	30%	30%	30%	20%	50%	10%	40%	22%	33%	22%	44%	0%	44%	13%	63%	27%	63%		

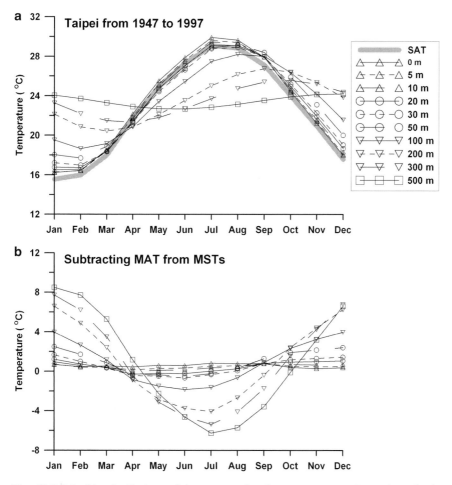

Fig. 10.3 Monthly distributions of long-term subsurface temperatures for various depths (0–500 cm) (MSTs) and MAT, and the patterns of their difference. MSTs and MAT given by averages of the entire observation period of data in each month are shown in (**a**). The patterns of MAT subtracting from MSTs are displayed in (**b**)

to their latitudes. The averages and ranges of the differences at the shallow depths are both inversely proportional to the latitude (Fig. 10.4a, b). The positive averages mainly ranged between 0°C and 3°C clearly indicate temperature discontinuity does in existence near the Earth surface due to distinct thermal conductivity. With respect to deep depths, the averages of temperature difference gradually decrease with the higher latitude, and are in good agreement with those in shallow depths (Fig. 10.5a). The ranges of the difference, which are proportional to the latitude, imply that impacts of SAT are sharply reduced with depths (Fig. 10.5b). Furthermore, the difference range of the Jiyuehtan station with a relative high

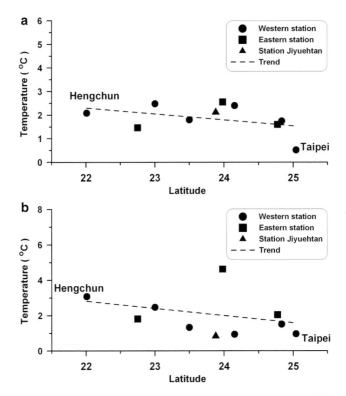

Fig. 10.4 Relationship between the averages and ranges of the differences (MSTs-MAT) at the shallow depths (<10 cm) with latitudes. Averages and ranges of the differences are shown in (**a**) and (**b**), respectively. *Circles* denote the stations located in the western Taiwan. In contrast, *squares* indicate the stations located in the eastern Taiwan. Station Jiyuehtan with the relative high altitude is marked by *triangle symbols*. Note that the *dashed lines* show the linear trends of the relationship

altitude is significantly smaller than those of other stations of similar latitude due to small variations of MAT. By comparing the stations with similar latitudes (Hsinchu V.S. Ilan, Taichung V.S. Hualien and Tainan V.S. Taitung), it is interesting to note that the difference ranges of the western stations are generally larger than those of the eastern stations with a range of about 1–3°C.

To study the departing of extreme temperatures of July and January relative to depth, the subsurface temperature at 0 cm were taken as a reference to estimate the time shift for each observation depth using the cross correlation method (Campbell et al. 1997). The time shifts of temperature at each depth relative to surface in all stations are listed in Table 10.3, and the relationships of the Hualien and Ilan stations are shown in Fig. 10.6 in parallel. The time shifts are shorter than 3 days at all stations with a depth shallower than 50 cm (Table 10.3 and Fig. 10.6). When the

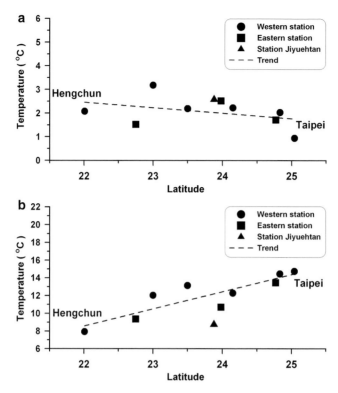

Fig. 10.5 Relationship between the averages and ranges of the differences (MSTs-MAT) at the deep depths (>100 cm) with latitudes. Averages and ranges of the differences are shown in (**a**) and (**b**), respectively. *Circles* denote the stations located in the western Taiwan. In contrast, *squares* indicate the stations located in the eastern Taiwan. Station Jiyuehtan with relative high altitude is marked by *triangle symbols*. *Dashed lines* show the linear trends of the relationship

Table 10.3 The time shifts relative to peaks of ground surface temperature each depth in the stations (unit: day)

Station	5 cm	10 cm	20 cm	30 cm	50 cm	100 cm	200 cm	300 cm	500 cm
Taipei	1	1	2	2	3	21	52.5	74.5	134
Hsinchu	1	1	2	2	3	15	41	53.5	91
Taichung	1	1	2	2	3	12	40	58	83
Jiyuehtan	1	1	2	2	3	18.5	43	54	82
Chiayi	1	1	2	2	3	23	45	72	120
Tainan	1	1	2	2	2	13	47	68.5	104
Hengchun	1	1	2	2	3	7	43.5	69	116.5
Taitung	1	1	2	2	2	5	42	55	85
Hualien	1	1	1.5	2	3	5	29	48	85
Ilan	1	1	2	2	2	18	44	70	120

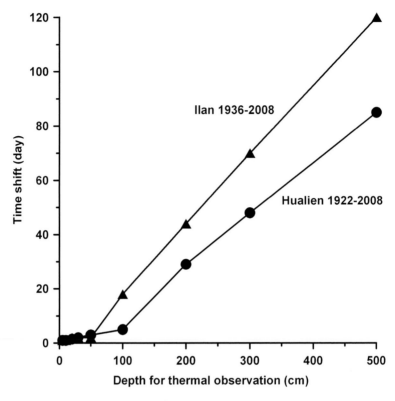

Fig. 10.6 Relationship between time shifts with various depths. Lines with *triangles* and *circles* denote the relationships in Ilian and Huanlien, respectively

depth gets deeper than 100 cm, the time shifts apparently increase with a linear trend and reach 82–134 days at the 500 cm depth (Table 10.3).

10.3   Discussion and Conclusions

The time shifts from 0 to 500 cm can be separated into three groups (Table 10.3). The short time shift (82–85 days) at the Jiyuehtan, Taichung, Taitung and Hualien stations infers that relatively high efficient geothermal conductivity is mainly distributed in the eastern part of Taiwan and extends to the western side of central Taiwan. Stations Hsinchu and Tainan with the medium time shift (91 and 104 days, respectively) are located at two transitional zones. Two end-member areas, which are either north or south from Hsinchu and Tainan, appear to have the longer time shift (116.5–134 days). The long time shift suggests that heat takes much time for propagating from the shallow to deep depth at the Hengchun, Chiayi, Ilan and Taipei stations. The variations of time shifts among stations indicate that the geothermal conductivity is an important factor even if temperature history is reconstructed in a small region.

Stations with long temporal periods and evenly distribution in Taiwan generate good temperature records to study the relationship between GST and SAT for various environment conditions. High SAT with small variations is observed in the low latitude and tropical climate. On contrast, low SAT with large variations is recorded in the high latitude. Figure 10.4 shows the ranges and averages of the differences between MST and MAT of shallow depths are both inversely proportional to the latitudes. For stations that are located in the low latitude with high SAT, temperature yields large differences between surface and subsurface records. In terms of the deep depths, the ranges of the difference (MST-MAT) are increased with higher latitudes (Fig. 10.5b). This implies that depth effect of high SAT is smaller than those of low SAT. Thus, the depth effect needs to be taken into account during temperature reconstructions processes.

For site comparison, the yearly mean of SAT presents an increase trend along with global warming and is consistent with most areas in the world. On the other hand, the subsurface temperature with decrease trends is observed at most depths in Taipei (Table 10.2). A further comparison of the linear trends of similar latitude stations (Hsinchu V.S. Ilan, Taichung V.S. Hualien and Tainan V.S. Taitung), it is clear to note that the temperature increase rate in eastern Taiwan is higher than it in the western part (Table 10.2). Because urban cities and rural sections can be roughly separated into the western and eastern sides of Taiwan, subsurface temperature with the small increase rates in the western side suggesting that high-rise buildings possibly cut off and/or absorb the inward thermal propagation and take hold of heat on the Earth surface, especially the holistic negative trends in metropolitan Taipei because Taipei has been the rapidest and most extensive developing area in Taiwan (Chen et al. 2007). These features imply that global warming signals may be underestimated due to contributions from negative feedbacks.

Furthermore, Table 10.2 exhibits that SAT generally has an increase trend during the entire observation period. However, subsurface temperatures with positive or negative changes can be found at various depths after 1980, but this feature is mainly confined in shallow depths before 1980. This suggests that the original patterns of GST and SAT are certainly different. In addition, the significant positive transformations before and after 1980 are widely observed both in subsurface and SAT in Taiwan. Therefore, a high correlation between GST and SAT should be dominated by the similar responses resulted from same impact sources. Temperature changes at different layers are roughly consistent among most stations. The temperature reconstructions should be generated using consistent transformation changes, correlated with the long-term trends at each depth and simultaneously take the time shift with depth into account. Meanwhile, global warming has persistently affected the subsurface temperature to generate significant positive changes on the trends after 1980. This is a serious warning that the persistent warming environments are not only affect the air and water domains, but also gradually extend to soil, rocks of the Earth's lithosphere.

References

Baker JM, Baker DG (2002) Long-term ground heat flux and heat storage at a mid-latitude site. Clim Change 54:295–303

Baker DG, Ruschy DL (1993) The recent warming in eastern Minnesota shown by ground temperatures. Geophys Res Lett 20:371–374

Beltrami H (2001a) Surface heat flux histories from geothermal data: inference from inversion. Geophys Res Lett 28:655–658

Beltrami H (2001b) On the relationship between ground temperature histories and meteorological records: a report on the Pomquet station. Glob Planet Change 29:327–352

Beltrami H (2002) Climate from borehole data: energy fluxes and temperatures since 1500. Geophys Res Lett. doi:10.1029/2002GL015702

Beltrami H, Kellman L (2003) An examination of short- and long-term air–ground temperature coupling. Glob Planet Change 38:291–303

Campbell JY, Lo A, MacKinlay AC (1997) The econometrics of financial markets. Princeton University Press, Princeton

Chen TC, Wang SY, Yen MC (2007) Enhancement of afternoon thunderstorm activity by urbanization in a valley: Taipei. J Appl Meteorol Climat 46:1324–1340

Harris RN, Chapman DS (2001) Mid latitude (30 deg – 60degN) climatic warming inferred by combining borehole temperature with surface air temperature. Geophys Res Lett 28:747–750

Ho CS (1988) An introduction to the geology of Taiwan, 2nd edn. Central Geological Survey, The Ministry of Economic Affairs, Taipei

Huang S, Pollack HN, Shen PY (2000) Temperature trends over the last five centuries reconstructed from borehole temperatures. Nature 403:756–758

IPCC (2007) Climate change 2007: the physical science basis – summary for policymakers, intergovernmental panel on climate change, Geneva, IPCC WGI 4th Assessment Report

Lloyd J, Taylor JA (1994) On the temperature dependence of soil respiration. Funct Ecol 8:315–323

Luo Y, Wan S, Hui D, Wallace L (2001) Acclimatization of soil respiration to warming in a tall grass prairie. Nature 413:622–625

Pollack HN, Huang S (2000) Climate reconstructions from subsurface temperatures. Annu Rev Earth Planet Sci 28:339–365

Putnam SN, Chapman DS (1996) A geothermal climate change observatory: first tear results from Emigrant Pass in northwest Utah. J Geophys Res 101:21877–21890

Rao CR, Toutenburg H, Fieger A, Heumann C, Nittner T, Scheid S (1999) Linear models: least squares and alternatives, 2nd edn. Springer Verlag, Berlin, Heidelberg, New York (Springer Series in Statistics)

Risk D, Kellman L, Beltrami H (2002a) Carbon dioxide in soil profiles: production and temperature dependence. Geophys Res Lett. doi:10.1029/2001GL014002

Risk D, Kellman L, Beltrami H (2002b) Soil CO_2 production and surface flux at four climate observatories in eastern Canada. Glob Biogeochem Cycles. doi:10.1029/2001GB001831

Smerdon JE, Pollack HN, Enz JW, Lewis MJ (2003) Conduction-dominated heat transport of the annual temperature signal in soil. J Geophys Res. doi:10.1029/2002JB002351

Smerdon JE, Pollack HN, Cermak V, Enz JW, Kresl M, Safanda J, Wehmiller JF (2006) Daily, seasonal and annual relationships between air and surface temperatures. J Geophys Res. doi:10.1029/2004JD005578

Wolberg J (2005) Data analysis using the method of least squares: extracting the most information from experiments. Springer Verlag, Berlin, Heidelberg, New York

Zhang T, Barry RG, Gilichinsky D, Bykhovets SS, Sorokovikov VA, Ye JP (2001) Am amplified signal of climatic change in soil temperatures during the last century at Irkutsk, Russia. Clim Change 49:41–76

Chapter 11
Evolution of the Subsurface Thermal Environment in Urban Areas: Studies in Large Cities in East Asia

Makoto Yamano

Abstract Temporal variation in the ground surface temperature (GST) propagates downward and disturbs the subsurface temperature structure. Analysis of subsurface temperature profiles gives information on the past GST history. For investigation of subsurface thermal environment evolution in urban areas, we reconstructed GST history from borehole temperature data obtained in large cities in East Asia. Most of the estimated GST histories show significant surface warming in the last century. In the Bangkok area, the amount of the GST increase is larger in the city center than in suburban and rural areas, corresponding to the degree of urbanization. The amount of heat accumulated in the subsurface due to surface warming, which can be calculated at the sites where GST histories were reconstructed, is a useful indicator of the subsurface thermal environment. We conducted monitoring of borehole temperature and soil temperature aiming to observe process of downward propagation of GST variation. Data obtained in some wells exhibit periodic or long-term temperature variations attributable to human activities. Soil temperatures measured within 1 m of the ground surface show prominent annual variation and provide information on the relation between GST and air temperature.

11.1 Introduction

Temperature in the underground generally increases with depth and heat flows upward toward the ground surface. Density of this heat flux at the earth's surface is termed "terrestrial heat flow" (or simply "heat flow"). Heat flow is defined as the amount of heat flowing out through the ground surface per unit time and unit area. It is practically determined as the product of the vertical temperature gradient near the surface and the thermal conductivity of the formations.

M. Yamano (✉)
Earthquake Research Institute, University of Tokyo, 1-1-1 Yayoi, Bunkyo-ku,
Tokyo 113-0032, Japan
e-mail: yamano@eri.u-tokyo.ac.jp

M. Taniguchi (ed.), *Groundwater and Subsurface Environments: Human Impacts in Asian Coastal Cities*, DOI 10.1007/978-4-431-53904-9_11, © Springer 2011

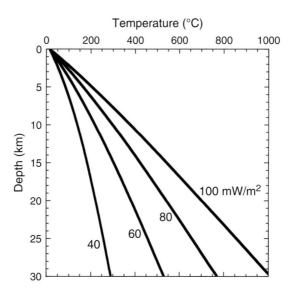

Fig. 11.1 Temperature versus depth curves corresponding to surface heat flow of 40, 60, 80 and 100 mW/m² calculated with typical thermal property values for continental crust (from Yamano 2010)

Since we cannot directly measure temperature deep in the crust, we need to estimate temperature in the subsurface from some data which can be obtained at around the earth's surface. Terrestrial heat flow is the most fundamental and useful observable for this purpose. Higher heat flow at the surface generally indicates higher temperature in the underground (Fig. 11.1). The subsurface temperature structure can be estimated based on the heat flow distribution at the surface.

Although subsurface temperature distribution is essentially determined by the heat flow from the deep in the crust, the temperature in the vicinity of the surface is disturbed by various phenomena such as groundwater flow and erosion/sedimentation. One of the major sources of the disturbance is temporal variation in the ground surface temperature (GST). GST variation propagates through subsurface formations by thermal diffusion. In Fig. 11.2, a temperature profile measured in a borehole in 1970 is compared with the profiles calculated assuming GST increase from 1880 to 1940 (Jessop 1990). The calculated profiles agree well with the observation, showing that the prominent curvature of the measured profile is a result of GST increase.

The disturbance in temperature profiles by the GST variation is noise for heat flow measurements and estimation of the thermal structure in deep in the crust and upper mantle. From a different point of view, it can be considered as a signal which gives information on GST variation in the past. Analysis of borehole temperature profiles should allow us to estimate GST variation history in the past. Studies of the past climate change based on this principle have been extensively conducted since the 1980s mainly in North America and Europe and made significant contribution

Fig. 11.2 Temperature profiles in a borehole in Canada: calculated for linear rise in GST (ground surface temperature) from 3.2°C in 1880 to 5.4°C in 1940 (*solid curves*) and measured in 1970 (*crosses*) (from Jessop 1990)

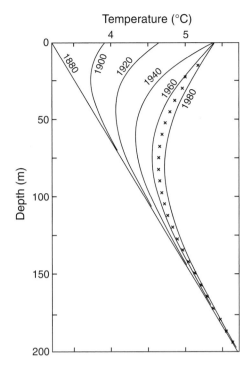

to investigation of recent global warming (e.g., Harris and Chapman 1997; Huang et al. 2000; Pollack and Huang 2000).

One of the most principal causes of GST change is variation in the surface air temperature (SAT). Long-period SAT variations, such as recent global warming and the Little Ice Age, have significant effect on the subsurface temperature distribution, whereas influence of shorter-period variation (diurnal or annual) is fully attenuated at shallow depth. In addition to global climate changes, local SAT changes, including the heat island effect due to urbanization, may have marked influence on the subsurface thermal environment.

GST is influenced not only by SAT but also by changes in ground surface environment (e.g., Lewis and Wang 1992; Bartlett et al. 2004; Woodbury et al. 2009). Difference in land coverage (e.g., bare rock, grassland, forest, or city) should yield different annual mean GST for the same annual mean SAT. Snow cover acts as a thermal insulator and gives a bias in GST and SAT relation in winter. GST variation in the past estimated through analysis of the subsurface temperature distribution, therefore, contains information on ground surface environment as well as SAT.

Human activities associated with urbanization lead to GST variation due to development of "heat islands", land use change and other environment changes. History of GST variation caused by such human activities must have been recorded in subsurface temperature distribution in urban areas. We have been studying evolution of thermal environment around the ground surface in large cities in East Asia through measurement and analysis of temperature profiles in boreholes.

11.2   Downward Propagation of GST Variation

Influence of temporal variation of GST propagates downward through the subsurface material by thermal diffusion. Propagation process of temperature disturbance by some basic patterns of GST change is discussed in this section.

For simplicity, it is assumed that subsurface material is uniform and the temperature structure is one-dimensional (in the vertical direction only). Temporal variation of the subsurface temperature distribution is described by the equation of thermal diffusion:

$$\frac{\partial T}{\partial t} = \kappa \frac{\partial^2 T}{\partial z^2} \tag{11.1}$$

where T is temperature, t is time, z is depth, and κ is thermal diffusivity.

We first examine a case where the GST is a sinusoidal function of time with a period of P and an amplitude of A:

$$T(0,t) = A\cos\left(\frac{2\pi t}{P}\right) \tag{11.2}$$

By solving (11.1) with the boundary condition (11.2), we obtain the temperature disturbance in the subsurface:

$$T(z,t) = A\exp\left(-\sqrt{\frac{\pi}{\kappa P}}z\right)\cos\left(\frac{2\pi t}{P} - \sqrt{\frac{\pi}{\kappa P}}z\right) \tag{11.3}$$

The amplitude of temperature variation is exponentially attenuated and the phase is linearly shifted with depth. We may define a characteristic penetration depth of temperature disturbance, z_p, as the depth at which the amplitude of temperature variation is $1/e^2$ (about 14%) of that at the surface, that is $z_p = 2\sqrt{\kappa P / \pi}$. Assuming that κ is 1×10^{-6} m²/s, z_p is about 6.3 m for a period of 1 year, which means that temperature disturbance by annual GST variation is confined in very shallow formations. Influence of longer-period variation propagates deeper (z_p is proportional to the square root of the period). For a period of 200 years, z_p is about 90 m and the disturbance reaches beyond 100 m (Fig. 11.3).

Another simple case is a step-function change in GST. If we assume a sudden increase in GST of ΔT at $t=0$:

$$\begin{aligned} T(0,t) &= 0 \quad & t < 0 \\ &= \Delta T \quad & t \geq 0 \end{aligned} \tag{11.4}$$

the temperature disturbance in the subsurface is:

$$T(z,t) = \Delta T\left[1 - erf\left(\frac{z}{2\sqrt{\kappa t}}\right)\right] \tag{11.5}$$

where erf (x) is the error function of x. Penetration of the effect of step-function GST change into the subsurface is demonstrated in Fig. 11.4. Equation (11.5)

Fig. 11.3 Subsurface
temperature perturbation by a
sinusoidal GST change with
a unit amplitude and a period
of 200 years (every 25 years)

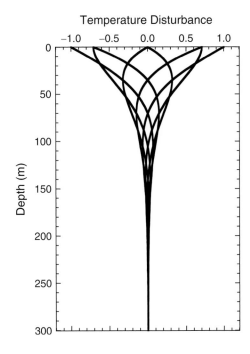

Fig. 11.4 Subsurface
temperature perturbation by a
sudden increase in GST by
unit temperature (1, 10 100,
and 500 years after the
increase)

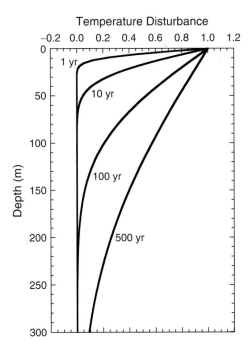

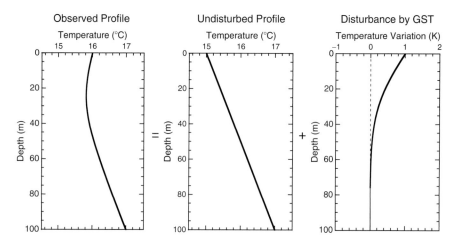

Fig. 11.5 Schematic temperature profile disturbed by GST variation, which can be decomposed into the undisturbed profile and the disturbance by GST variation

shows that the depth of penetration of temperature disturbance is proportional to the square root of the time since the GST change. If we consider a characteristic penetration depth $z_s = 2\sqrt{\kappa t}$, at which the temperature change is about 16% of ΔT, z_s 100 years after the GST change is about 110 m for κ of 1×10^{-6} m²/s. Penetration depth (both z_s and z_p) depends on thermal diffusivity of the subsurface material (soil or rock), which generally range from 3×10^{-7} to 2×10^{-6} m²/s. As can be seen in Fig. 11.4, information on the GST variation in the last several hundred years has been recorded in the upper several hundred meters in the underground.

The actual GST variations can be approximated by a sum of sinusoidal functions with various periods or a series of step functions. Temperature disturbance by the GST variations can then be easily calculated by superposing the solutions (11.3) or (11.5). The observed subsurface temperature distribution is the sum of the temperature disturbance and the undisturbed profile determined by heat flow from the deep and thermal conductivity (Fig. 11.5).

11.3   Reconstruction of GST History in the Past

Temporal variation of GST propagates downward by thermal diffusion and influences the underground temperature structure, as shown above. Information on the past GST variation can therefore be extracted from the disturbed subsurface temperature distribution. With this "geothermal" method, we can reconstruct GST history in the past from temperature profiles measured in boreholes.

Supposing that the past GST variation is approximated by a sum of a series of step function (Fig. 11.6), the influence of the GST variation on the subsurface temperature structure can be calculated using (11.5). It is also possible to estimate GST history in the past which best reproduce the observed subsurface temperature profiles. This estimation analysis is usually made using inversion formulations. Model

parameters to be determined by inversion are GST change for each step (difference
from the reference temperature; cf. Fig. 11.6), thermal diffusivity, and the undisturbed temperature gradient corresponding to heat flow from the deep. In cases
where GST variations are expressed as superposition of sinusoidal oscillations,
similar inversion analysis based on (11.3) determines amplitude and initial phase of
each oscillation component.

GST history reconstruction studies using the "geothermal method" started
around 1970 and have been extensively conducted since the late 1980s for the purpose of investigating climate change in the last several hundred years (e.g., Cermak
1971; Lachenbruch and Marshall 1986; Wang 1992; Pollack and Chapman 1993).
Analysis techniques for reconstruction of GST from borehole temperature profiles
have been well established through a number of theoretical studies (cf. Pollack and
Huang 2000).

Estimation of GST history has been made mainly in the North American and
European continents. Figure 11.7 shows GST variations estimated from temperature

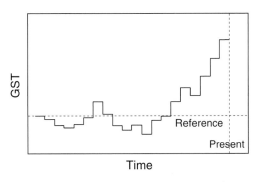

Fig. 11.6 Expression of
GST variation in the past as a
series of step function

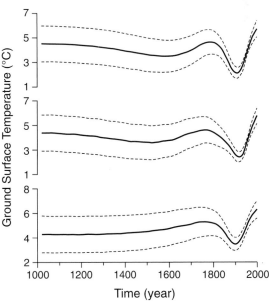

Fig. 11.7 GST histories for
the last 1,000 years estimated
through analysis of
temperature profiles
measured in three boreholes
in Quebec, Canada (from
Wang and Lewis 1992)

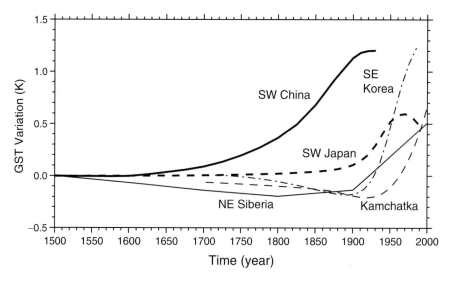

Fig. 11.8 GST histories reconstructed from borehole temperature profiles in five areas of East Asia (Huang et al. 1995; Pollack et al. 2003; Goto et al. 2005, 2009; Cermak et al. 2006) (from Yamano et al. 2009)

profiles measured in three closely-located boreholes in Quebec, Canada by inversion analysis (Wang and Lewis 1992). The GST histories reconstructed at the three holes, which are located within 2 km, are similar to one another, suggesting that reconstruction analysis was successful at this site. They show the existence of a cold period around 1900 before the start of surface warming over the twentieth century.

In Asia, such borehole climate study was conducted only in limited areas (e.g., in Russia, China and India). Among them, results of GST reconstruction in five areas in East Asia (Huang et al. 1995; Pollack et al. 2003; Goto et al. 2005, 2009; Cermak et al. 2006) were compiled in Fig. 11.8. All the GST histories, reconstructed in widely distributed areas such as NE Siberia, Kamchatka, and SW China, show significant surface warming in the last 100–200 years. The amount of temperature increase and the onset time of warming are, however, different from each other. The GST histories should reflect various factors, including global warming, local climate changes, and influence of urbanization, which may have resulted in the differences among the areas.

As compared with other methods for paleo-climate reconstruction, the past GST estimation from borehole temperature data has the following features: (1) GST history can be directly estimated from subsurface temperature distribution, not through conversion from proxies related to paleo-SAT. It should be noted, however, that GST does not generally agree with SAT. (2) We can obtain the long-term trend of temperature variation with time scales of several 10 years or longer, while yearly variation cannot be resolved. This feature arises from characteristics of thermal diffusion that temperature variation signal diffuses and attenuates as it propagates downward through subsurface formations. (3) Times for which no meteorological data are available, including those before industrialization, can be covered. This method is therefore effective in studies of impacts of human activity on surface thermal environment.

11.4 Measurements of Temperature Profiles in Boreholes

We applied the geothermal method of GST history reconstruction in studies on evolution of the thermal environment of large cities in East Asia (Yamano et al. 2009) as a part of an international multidisciplinary research project "Human Impacts on Urban Subsurface Environments" by the Research Institute for Humanity and Nature (Taniguchi et al. 2009). GST histories obtained in areas in various stages of development from the city center to surrounding rural areas would provide information on formation of a "heat island" associated with development of the city. Comparison of the results in the city with those in rural areas may enable us to differentiate the effect of heat island from that of global warming.

We first need to obtain subsurface temperature profile data for GST history reconstruction analysis. In our target cities, Tokyo, Osaka, Seoul, Taipei, Bangkok, and Jakarta, a number of observation wells have been drilled for groundwater monitoring for investigation of land subsidence and water resource problems. Most of the wells are suitable for temperature measurement, because they are used for passive monitoring and not disturbed by pumping and other activities. Basic geological and hydrogeological information on the surrounding formations are also available. We thus made temperature profile measurements mainly in these groundwater monitoring wells.

In the Tokyo and Osaka areas, temperature profile measurements had been made at many stations for the purpose of investigation of groundwater flow and subsurface thermal environment (e.g., Dapaah-Siakwan and Kayane 1995; Taniguchi and Uemura 2005). Consequently, temperature measurement surveys were started in Seoul, Taipei, Bangkok, and Jakarta, where few temperature profile data were available (Table 11.1; Hamamoto et al. 2009; Lubis et al. 2009).

We lowered a temperature sensor in observation wells and measured temperature at depth intervals of 1 or 2 m. Water inside the well is considered to be in thermal equilibrium with the surrounding formation, if there is no convection of borehole water. Temperature above the water level is much disturbed by convection of air, which sets the upper depth limit of temperature profile measurement at the water level.

In Seoul, temperature profile measurements were made at 14 stations. All of the measured holes were shallower than 90 m. The temperature profiles have been severely disturbed by pumping and/or groundwater flow and cannot be used for analysis. In the Bangkok, Taipei, and Jakarta areas, we conducted temperature measurement surveys four times. Survey time, number of stations where profile measurements were made, and measurement depths are listed in Table 11.1. In Taiwan, measurements have been made in the southern part, in the vicinities of Chiayi, Tainan and Pingtung, in addition to the area in and around the city of Taipei.

Table 11.1 Borehole temperature measurements in and around large cities in East Asia

City	Survey time	Number of stations[a]	Measurement depth[b]
Seoul	Sep. 2005	14	14–88 m
Bangkok	Jul. 2004	27	40–401 m
	Jun. 2006	19	
	Mar. 2008	16	
	Feb. 2010	9	
		Total 44	
Taipei[c]	Nov. 2005	11	60–308 m
	Jun. 2007	18	
	Jan. 2009	14	
	Feb. 2010	5	
		Total 26	
Jakarta	Sep. 2006	28	40–252 m
	Aug. 2007	12	
	Aug. 2008	6	
	Feb. 2009	2	
		Total 28	

[a] Total number is less than the sum of the numbers of each survey because of repeated measurements
[b] Range of the maximum measurement depth in each well
[c] Including the southern and southwestern parts of Taiwan

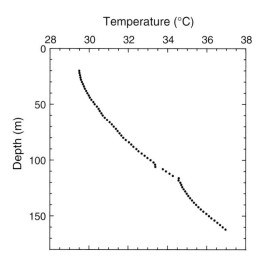

Fig. 11.9 Borehole temperature profile measured in the Jakarta area, which shows a bend probably due to groundwater flow (from Yamano 2010)

GST reconstruction analysis is made assuming that heat transfer in the subsurface is practically conductive and advection of heat by groundwater flow is negligible. It is hence necessary to discard temperature profile data affected by groundwater flow. Influence of groundwater flow is obvious in some of the profiles, which are significantly distorted. Figure 11.9 shows a typical disturbed temperature profile, which has a peculiar bend around 115 m. Even if a temperature profile is not apparently

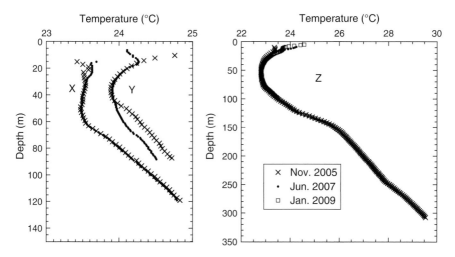

Fig. 11.10 Temperature-depth profiles in three boreholes (X, Y and Z) in the Taipei area measured in November 2005 (*crosses*), June 2007 (*dots*), and January 2009 (*squares*)

distorted, it might be affected by groundwater flow. A possible way to detect such disturbed data is examination of the stability of temperature profiles through repeated measurements.

Temperature profiles measured repeatedly at three stations in the Taipei area are shown in Fig. 11.10. Although the shapes of the temperature profiles in wells X and Y appear to be similar to each other, the stability of the profile was very different. In well Y, the profile measured in November 2005 significantly different from the one in June 2007 below 45 m, while the profile in well X did not show any appreciable change. The instability of the profile in well Y may be due to temporal change in groundwater flow around it. It should be noted that stable profiles could also be affected by stable groundwater flow. In well Z (Fig. 11.10), the temperature gradient considerably varies with depth around 120–160 m, indicating influence of groundwater flow, but the temperature profile had been stable through three measurements in November 2005, June 2007 and January 2009.

11.5 GST Histories Reconstructed in the Bangkok Area

Subsurface temperature profiles measured in boreholes in the target city areas provide information on evolution of the thermal environment around the ground surface. Most of the profiles showed small (or even negative) temperature gradients in the upper parts of the holes, as can be seen in Fig. 11.11 (examples of profiles obtained in the Bangkok area and in southern Taiwan). Similar temperature profiles were reported in and around Tokyo and Osaka as well (e.g., Taniguchi et al. 1999; Taniguchi and Uemura 2005). Such curvature of the profiles is most probably

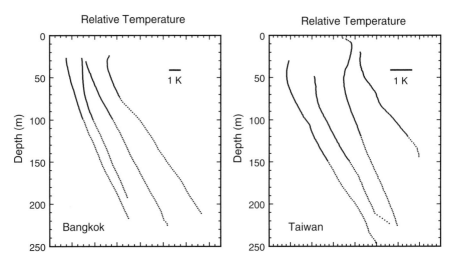

Fig. 11.11 Examples of temperature profiles measured in observation wells in the Bangkok area in June 2006 (**a**) and in southern Taiwan in January 2009 (**b**) (from Yamano 2010). The profiles are shifted on the temperature axis so that they do not overlap with each other

attributable to recent increase in GST. We should, however, consider the possibility that subsurface temperature distribution is seriously perturbed by groundwater flow.

Downward groundwater flow, prevailing in recharge areas of regional flow systems, advectively transports heat downward and makes temperature profiles concave toward the surface (e.g., Domenico and Palciauskas 1973). Similarly, upward flow in discharge areas results in temperature profiles convex toward the surface. If the flow is steady, uniform, and vertical, it is possible to estimate the Darcy velocity and the total heat flux (both conduction and advection) from the shape of the temperature profile (Bredehoeft and Papadopulos 1965).

Taniguchi et al. (1999) compared borehole temperature profiles observed in the Tokyo area with a subsurface thermal model in which GST increase in the last century and advective heat transfer by vertical groundwater flow are incorporated. They found that the penetration depths of the influence of surface warming are deeper in the uplands in the west and shallower in the lowlands in the east. The difference is interpreted to result from downward groundwater flow in the uplands (recharge area) and upward flow in the lowlands (discharge area). Similar studies of influence of surface warming and groundwater flow system on the subsurface temperature distribution have been made in some other areas in Japan (e.g., Miyakoshi et al. 2003; Uchida et al. 2003).

In the areas where subsurface temperature is significantly affected by groundwater flow, averaging many temperature profiles over the area may reduce the effect of groundwater flow and allow us to extract information on GST variation from the temperature profiles. Taniguchi et al. (2007) compiled temperature profile data in and around Tokyo, Osaka, Seoul, and Bangkok and calculated the average profile for each city. They estimated the onset time and the amount of recent GST increase based on the average profiles. The result indicates that surface warming started earliest in Tokyo and latest in Bangkok, consistent with development histories of these cities, and the warming rate in each city roughly agrees with the increase rate of annual mean SAT. It suggests that the average of subsurface temperature profiles could represent the thermal environment evolution of the city to some extent. It is necessary, however, to analyze temperature profiles free from influence of groundwater flow for more quantitative discussion and for examination of variation of GST histories within a city area.

In the Bangkok area, GST history reconstruction analysis was made on selected temperature profiles (Hamamoto et al. 2009). Temperature profile measurements in the Bangkok area were made in groundwater monitoring wells at 44 sites (Table 11.1). Most of the wells are located in and around the city of Bangkok, while some wells are about 100 km north of Bangkok (Fig. 11.12). Surface environment around the observation wells varies largely from the downtown in Bangkok to rural agricultural land (Fig. 11.13). The obtained temperature profiles were examined on the shape and stability to discard ones affected by groundwater flow, and six sites were chosen as suitable for GST history reconstruction analysis (Fig. 11.14).

GST histories at the six sites were estimated through inversion analysis of the temperature profiles using a multi-layer model (Goto and Yamano 2010). Thermal properties of subsurface material generally vary with depth depending mainly on

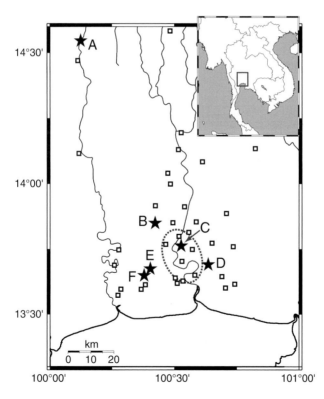

Fig. 11.12 Locations of groundwater monitoring sites in the Bangkok area where temperature profile measurement was conducted. Stars are the sites where GST reconstruction analysis was made (Hamamoto et al. 2009). The *broken ellipse* approximately shows the city of Bangkok

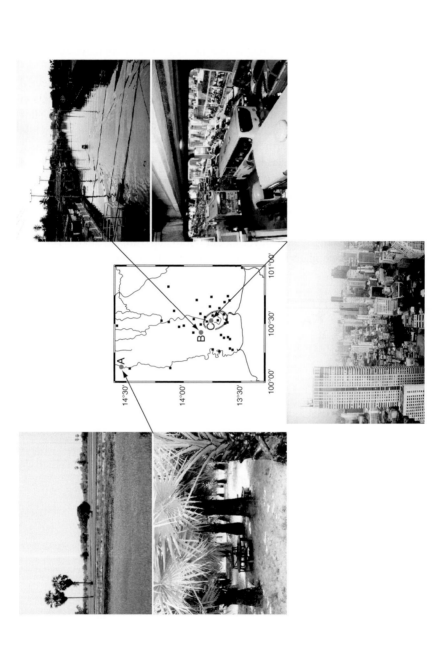

Fig. 11.13 Variety of environment around observation wells in the Bangkok area. Photographs on the *left* and on the *upper right* were taken in the vicinity of site A and site B respectively, while ones on the *lower right* and the *bottom* demonstrate typical environment in the city center

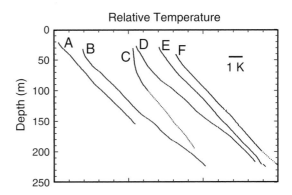

Fig. 11.14 Temperature profiles measured in March 2008 at the sites selected for GST reconstruction analysis in the Bangkok area (Hamamoto et al. 2009)

the porosity and mineral composition, while we assumed uniform thermal properties in the Sects. 11.2 and 11.3. Most of our target cities, including Bangkok, have been developed on alluvial plain, where subsurface formations may consist of alternating coarse and fine deposits. Formations around the observation wells are thus expected to have highly variable thermal properties. In the multi-layer model, thermal diffusivity of each layer is treated as a model parameter and the best-fitting value is estimated through inversion. Depths of the layer boundaries were determined based on available information on lithology around the wells.

The time span for which GST history can be reconstructed depends on the depth range of temperature profile measurement. Since the maximum depths of temperature measurements in the six wells are around 200 m, information on GST history for about 300 years may be extracted from the temperature profiles. The water levels in the wells, 20–40 m below the surface, set the upper limit of temperature profile data, preventing us from estimating GST in the last 10–20 years. Consequently, GST history reconstruction was conducted for 1700–1990 using a series of step functions with a step interval of 10 years (Fig. 11.15). It should be noted that GST estimated for each 10-year period is a kind of average temperature over a longer period especially at older times because of diffusion of temperature signal during the downward propagation process. Older signals have been more diffused, which may result in no appreciable variation in the estimated GST in the first 150 years.

In contrast, the estimated GST increased in the last 100–150 years at all the six sites (Fig. 11.15). The amount of temperature increase significantly varies by site. The temperature increase after 1900 is about 2.5 K at site C in the central part of Bangkok, while it is only about 0.4 K at site A in a rural area far from Bangkok (cf. Fig. 11.12). It suggests that the large GST increase at site C mainly arose from influence of urbanization and the effect of global warming is minor. The amount of temperature increase at the other four sites in suburbs of Bangkok is in between those at site C and at site A. Among them, the temperature increase at sites E and F located in more recently developed areas started later and the amount of increase is less. These results appear to be consistent with the development history of the city of Bangkok.

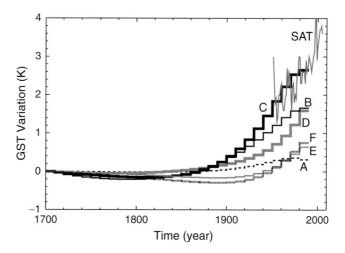

Fig. 11.15 GST histories reconstructed for the selected sites in the Bangkok area (Hamamoto et al. 2009) and mean annual surface air temperature at a meteorological station in Bangkok (Kataoka et al. 2009)

We need to discuss what processes in urbanization are responsible for the GST variations. Increase in SAT due to the development of "heat island" must have resulted in GST increase. Annual mean SAT at a meteorological station in Bangkok (Kataoka et al. 2009) has increased at a rate similar to that for GST at site C (Fig. 11.15). The SAT has been rising to the present, indicating that GST at site C has further increased after 1990.

GST, however, does not necessarily follow SAT closely (cf. Sect. 11.7). In the process of urbanization, land use change may also have an influence on GST. Changes in vegetation or land use generally affect coupling between SAT and GST (e.g., Lewis and Wang 1998; Beltrami and Kellman 2003). Ferguson and Woodbury (2007) investigated the subsurface temperature distribution in and around the city of Winnipeg, Canada and showed that temperature at 20 m depth conspicuously varies even within a few km and shows a clear correlation with land use (agricultural land, urban greenspaces, and urban areas). It is necessary to examine the relationship between GST variation and land use change in the Bangkok area as well.

We should also consider influence of the groundwater flow system on GST reconstruction analysis. Although temperature profiles without apparent disturbance by groundwater flow were used for analysis, there is still a possibility that the profiles have been affected by slow and steady flow. Groundwater flow system in the Bangkok area has been studied with hydrological and geochemical approaches and regional flow models have been presented (e.g., Yamanaka et al. 2009). The flow models can provide information on the direction and rate of groundwater flow around the wells where temperature profiles were measured. With this information, evolution of subsurface thermal environment may be simulated taking account of heat advection by groundwater flow.

Similar analysis of subsurface temperature profiles can be made in the Tokyo, Osaka, Jakarta, Taipei areas as well. Difference in GST increase between the city center and suburbs was also reported in a study in the Osaka area by Taniguchi and Uemura (2005). They compared the shapes of temperature profiles with minimal influence of groundwater flow and inferred that GST increased with an earlier onset time and by a larger amount at a site in the central part of Osaka than at sites in the surrounding areas. Further analysis of these and other profiles, including GST history reconstruction, would yield more quantitative results.

11.6   Accumulation of Heat Beneath the Large Cities

GST in urban areas has been increasing due to development of "heat island" associated with urbanization and/or global warming. Influence of GST increase propagates downward and causes a rise in subsurface temperature (Fig. 11.16). In other words, heat has been stored in the subsurface of urban areas by downward heat flux as a result of surface warming. The gray area between the temperature profile affected by surface warming and the undisturbed profile in Fig. 11.16 corresponds to accumulation of heat in the subsurface.

The amount of heat accumulated in the subsurface through this process has been estimated in a global scale. Beltrami et al. (2006) calculated surface heat flux histories from subsurface temperature profiles at 588 sites in the Northern Hemisphere and then estimated the average history over the hemisphere. On average, the Northern Hemisphere continental surface has absorbed heat since the beginning

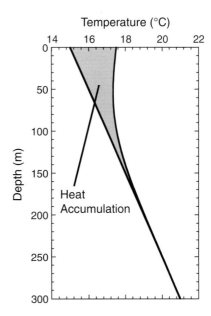

Fig. 11.16 Accumulation of heat in the subsurface (*gray area*) as a result of temperature increase caused by surface warming

of the nineteenth century and the heat flux was larger in the twentieth century. The total heat stored in the continental subsurface in the Northern Hemisphere between 1780 and 1980 was estimated to be 13.2×10^{21} J and 36% of the total (4.8×10^{21} J) was gained in the last 50 years (1930–1980).

Huang (2006) calculated the annual heat budget of each continent based on SAT data from 1851 to 2004, assuming that GST varied in parallel with SAT. The amount of heat absorbed by all the continents except for Antarctica in the 154 years was estimated as 11.7×10^{21} J. As for 50 years from 1951 to 2000, the heat content of the continents increased by 6.7×10^{21} J and that of Eurasia and North America increased by 4.6×10^{21} J, which is compatible with the heat gain in the Northern Hemisphere between 1930 and 1980 estimated by Beltrami et al. (2006), 4.8×10^{21} J. The consistent results by two different methods indicate that the estimated values are rather accurate.

We may estimate the amount of heat stored in the subsurface of urban areas, where significant GST increase occurred as a result of development of heat island by human activity. The heat stored in the subsurface per unit area till the time t since t_0 (reference time) can be calculated from temperature profiles at t and t_0 as:

$$Q(t) = \rho c \int \left[T(z,t) - T(z,t_0) \right] dz \qquad (11.6)$$

where ρc is heat capacity of subsurface material, which is assumed to be constant for simplicity.

Analysis for GST history reconstruction from subsurface temperature profiles gives estimate of the undisturbed temperature profile at the site before the reconstructed GST variation occurred. It is thus possible to calculate temperature profile at time t based on the undisturbed profile and the GST history. We can then obtain the heat stored in the subsurface using (11.6).

Hamamoto et al. (2009) estimated the heat stored after 1900 at the six sites in the Bangkok area where GST histories were reconstructed (A through F in Fig. 11.12) by the above method with typical values of thermal diffusivity and heat capacity (6.0×10^{-7} m²/s and 2.5 MJ/m³ respectively). The estimated increase in the heat content per unit area after 1900 is plotted every 10 years in Fig. 11.17. As heat accumulation in the subsurface is a result of increase in GST, temporal variation of the heat content is similar to that of GST (Fig. 11.15). The amount of stored heat is largest at site C in the central part of Bangkok, smallest at site A in a rural area, and intermediate in suburbs of Bangkok. Heat accumulation started earlier at sites C and D, corresponding to earlier onset time of GST increase at these sites. The results can be compared with the average over continents. Heat content per unit area averaged over Eurasia and North America increased by about 90 MJ/m² from 1900 to 2000 according to Huang (2006). At site C, about 200 MJ/m² has been stored in the subsurface, over the double of the continental average.

The above example demonstrates that a large amount of heat has been accumulated in the subsurface of urban areas. It means that subsurface heat island has been formed in response to development of heat island above the ground in urban areas (Ferguson and Woodbury 2007; Miyakoshi et al. 2009a). The subsurface heat island effect can be

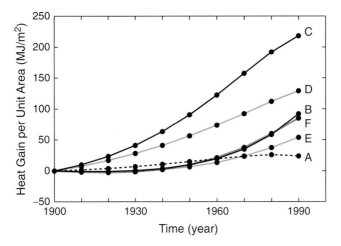

Fig. 11.17 Amount of heat stored in the subsurface after 1900 at the six sites in the Bangkok area (stars in Fig. 11.12) calculated based on the reconstructed GST histories (Hamamoto et al. 2009)

detected through repeated measurements of borehole temperature profiles with time intervals. In the Tokyo area, Miyakoshi et al. (2009b) compared the temperature profiles in observation wells measured in 2001–2002 and those measured in 2005–2006. They found that the subsurface temperature increased during this period and the rate of increase is higher in the central part of Tokyo than in suburban areas.

The amount of heat stored in the subsurface can be calculated in a straightforward way at sites where GST history reconstruction was made. It is a kind of depth integration of temperature anomaly in the subsurface and may serve as an indicator of the evolution of subsurface thermal environment in urban areas. It may be useful in multidisciplinary study on environmental changes in urban areas associated with city development.

11.7 Long-Term Temperature Monitoring

It is an important subject what information the past GST variation estimated from subsurface temperature profiles provides. GST histories obtained by the geothermal method are usually considered to represent climate change in the past and often used as basic data for studies on global warming. Such discussion is based on the assumption that reconstructed GST history is good approximation of SAT change on long timescales. Validity of this assumption has been investigated through numerical modelling and analysis of actual SAT and GST data. The results suggest that GST follows SAT change when averaged over a wide area on a timescale of 10 years (e.g., Gonzalez-Rouco et al. 2003; Smerdon et al. 2006). It is necessary, however, to examine the relationship between GST and SAT in detail for discussion on spatial and temporal variations of thermal environment of a city and its surrounding area.

The subsurface temperature distribution in the city of Winnipeg (Ferguson and Woodbury 2007; cf. Sect. 11.5) demonstrates that the GST and SAT relationship in an urban area is complicated. Huang et al. (2009) compared the temperature profiles measured in six boreholes in and around the city of Osaka with those calculated assuming that GST variation at these sites is approximated by SAT variation recorded at a meteorological station in the central part of Osaka. They found that the observed subsurface temperature disturbance by recent surface warming is larger than the calculated one at all the sites. It indicates that the rates of GST increase at these sites were higher than that of SAT at the Osaka meteorological station. Huang et al. (2009) suggested that the urban heat island effect might be more profound in the subsurface than in the air, since the heat transfer in the subsurface is dominated by conduction, which is much less efficient than air convection.

11.7.1 Monitoring of Soil Temperature

The GST estimated from subsurface temperature profiles, which is an average for a certain long period, can be compared with the annual mean SAT data, if long-time meteorological records are available in the vicinity. When there is no SAT data in the past, information on the present GST may be obtained through monitoring soil temperature just below the ground surface. The annual mean of temperature at a depth within 1 m of the surface must be close to the annual mean of GST. It is also possible to calculate GST from long-term records of temperatures at multiple depths (e.g., Smerdon et al. 2004; Bartlett et al. 2006).

We have been monitoring soil temperatures at shallow depths in the Bangkok, Jakarta, and Taipei areas aiming to estimate the present GST and to obtain information on the heat transfer process in surface soil. Temperature sensors were buried at two depths (about 0.5 and 1.0 m below the surface) beside the wells where temperature profiles were measured (Fig. 11.18). Soil temperature records for about 2 years at a site in Jakarta are shown in Fig. 11.19a. The temperature variation measured at 1.0 m is attenuated and phase shifted as compared to that at 0.5 m. They are the features of thermal diffusion process described by (11.3), which means we could directly observe downward propagation of GST variation through surface soil. Temperature variation at 1.0 m can be calculated from the temperature record at 0.5 m using (11.3) with an assumed average thermal diffusivity between the two depths. For appropriate values of thermal diffusivity, the calculated temperature agrees well with the observed one (Fig. 11.19a). It indicates that heat transfer between 0.5 and 1.0 m is almost conductive at this site. The best estimate of thermal diffusivity is 4.5 to 5.0×10^{-7} m^2/s.

Soil temperature records for over 2.5 years at a station in southern Taiwan can also be explained well by thermal diffusion process of the influence of GST variation (Fig. 11.19b). At this station, the average thermal diffusivity between 0.43 and 0.93 m below the ground surface is estimated to be about 8×10^{-7} m^2/s. These values are in a range of previously reported thermal diffusivity of shallow subsurface material (Smerdon et al. 2004; Bartlett et al. 2006), 4×10^{-7} to 1.0×10^{-6} m^2/s, determined

Fig. 11.18 Typical configuration of temperature monitoring systems installed in and around observation wells in the Bangkok, Jakarta and Taiwan areas (from Yamano 2010)

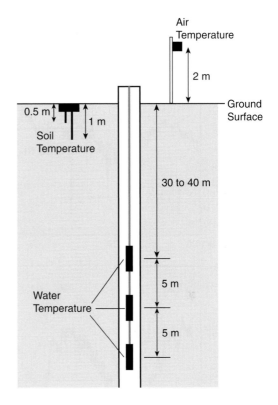

through analysis of temperature records. The variation of thermal diffusivity may be attributed to differences in the mineral composition, porosity, and degree of saturation.

At some of the sites where long-term monitoring of soil temperature has been conducted, SAT has also been monitored. They can be combined for analysis of the SAT and GST relationship.

11.7.2 Monitoring of Borehole Water Temperature

Long-term records of surface soil temperature give information on penetration process of GST variation with annual or shorter period as shown by the above example (Fig. 11.19). Propagation of longer period components of GST variation, which can be estimated through analysis of subsurface temperature profiles, may be observed by temperature monitoring at deeper depth. Cermak et al. (2000) conducted temperature monitoring experiments in boreholes in the Czech Republic and showed that temperature at 38 m below the ground surface monotonously increased by about 0.16 K in 6 years. This steady temperature increase is considered to be due to long-period components of GST variation, as short-period components including annual variation must have decayed before reaching this depth by thermal diffusion.

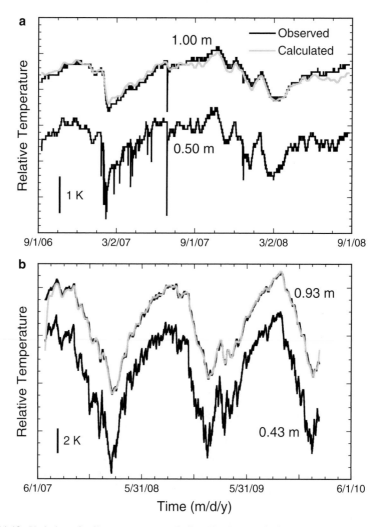

Fig. 11.19 Variation of soil temperatures at shallow depths at a site in Jakarta for 2 years (**a**) and at a site in southern Taiwan for 2.5 years (**b**). Theoretical temperature variation at deeper depths (*gray curves*, 1.00 m in (a) and 0.93 m in (b)) was calculated from the temperature records at shallower depths (0.50 m in (a) and 0.43 m in (b)) assuming pure thermal diffusion with diffusivity of 4.8×10^{-7} m^2/s (a) and 8.1×10^{-7} m^2/s (b)

Observation of downward propagation process of temperature variations allows us to examine the reliability of GST history reconstruction by the geothermal method, i.e. the validity of assumption of one-dimensional thermal diffusion in the vertical direction. We carried out temperature monitoring at 30–50 m depths in observation wells in the Bangkok, Jakarta, and Taipei areas. In each well, three self-contained water temperature recorders were installed at intervals of about 5 m (Fig. 11.18).

Temperature records for about 2.5 years obtained in a well in the Taipei area are shown in Fig. 11.20. Temperature increased monotonously at all the three depths (29.1, 34.2, and 39.4 m below the surface) except for two anomalous events in March 2008 and July 2009, and the rate of increase is higher at shallower depth. The temperature increase in this well may represent downward propagation of influence of recent surface warming as well as the one in the Czech borehole.

In another well in the Taipei area, peculiar short-period temperature oscillation was observed at one of three measurement depths, 25.0 m (Fig. 11.21a). A blowup of the 25.0 m record demonstrates that it contains 1-day and 1-week components (Fig. 11.21b), which is supported by the result of spectrum analysis. It strongly suggests that the short-period variation is related to some human activity. The water level in this well also oscillates with a 1-week component and has a negative correlation with the temperature at 25.0 m (Fig. 11.21b). Vertical movement of borehole water is therefore the most probable cause of the temperature variation. Temperature profiles measured in this well (Fig. 11.21c) show that the temperature gradient is negative at 25.0 m and nearly zero at the other two depths (33.2 and 41.4 m). It is consistent with the negative correlation between the temperature at 25.0 m and the water level and no significant temperature oscillation at 33.2 and 41.4 m. At this site, water level change due to human activity around the site (e.g., pumping of groundwater) must result in temperature variation in the well.

Temperature records in a borehole in the Lake Biwa Museum located on the coast of the Lake Biwa, Japan provide another example of influence of human activity around the ground surface (Yamano et al. 2009). Temperature distribution in the hole was measured repeatedly after completion of drilling in 1992. The profiles

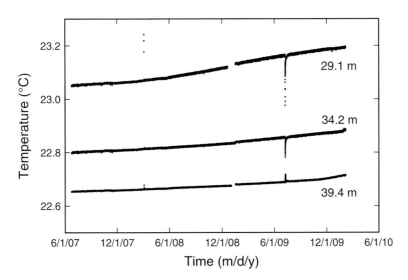

Fig. 11.20 Temperature records for about 2.5 years obtained at three depths in an observation well in the Taipei area

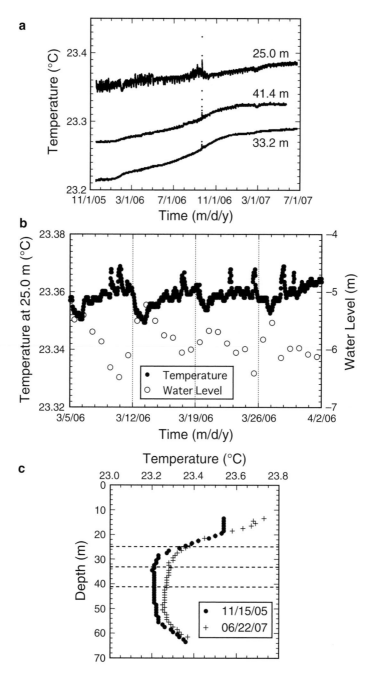

Fig. 11.21 (a) Temperature records at three depths in a groundwater monitoring well in the Taipei area. (b) Blowup of the temperature record at a depth of 25.0 m and the water level (provided by Institute of Earth Sciences, Academia Sinica, Taiwan). (c) Temperature profiles measured in the well in November 2005 and June 2007 (from Yamano 2010)

measured in 1993 and 2003 show that temperature above about 70 m depth significantly increased in the first 10 years after the drilling (Fig. 11.22a). It indicates that
the subsurface temperature structure at this site was disturbed by some recent
event(s) around the ground surface.

For investigation of this phenomenon, long-term temperature monitoring was
conducted at depths of 30 and 40 m. It was revealed that temperature had increased

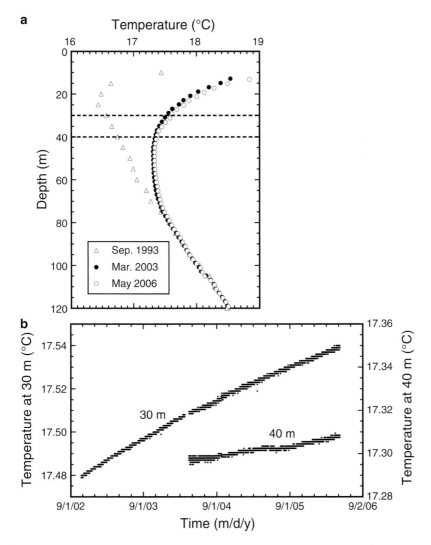

Fig. 11.22 (**a**) Temperature profiles in the upper part of a borehole in the Lake Biwa Museum,
Japan measured in September 1993, March 2003, and May 2006. *Dashed lines* are the depths at
which long-term temperature monitoring was conducted. (**b**) Temperature records at depths of 30
and 40 m (from Yamano et al. 2009)

at nearly constant rates at both depths (Fig. 11.22b). Probable causes of the temperature increases are: (1) an increase in the annual mean GST due to construction of the museum in 1994, which covered the ground surface around the hole or (2) an increase in the depth from the surface due to fill-up of artificial sediment (6.7 m thick) on the original ground surface sometime between 1982 and 1991. In the former model, a sudden increase in GST in 1994 would result in almost linear temperature increases at both depths during the monitoring period. The amount of GST increase corresponding to the increasing rate at 30 m, however, gives a much higher rate at 40 m than the observed one. The latter model also gives similar results and cannot explain the much larger temperature increase at 30 m than that at 40 m. We started temperature monitoring at ten depths (15–130 m) in 2007 to obtain more detailed information on the subsurface thermal process at this site.

Influence of artificial change in the surface environment on the subsurface temperature structure was detected through repeated measurements of temperature profiles at long time intervals as well. Safanda et al. (2007) analyzed temperature profiles repeatedly measured in a borehole in Prague and showed that construction of infrastructures on the ground surface around the site may have caused anomalous subsurface temperature increase. Ferguson and Woodbury (2007) found cooling in the subsurface above 50 m depth in a well in Winnipeg and suggested that it might be attributed to land use change from buildings to grass. These instances demonstrate that information on local environment changes at the ground surface is recorded in subsurface temperature distribution. Such local effects need to be considered carefully in interpretation of the GST history reconstructed from subsurface temperature profiles.

11.8 Summary

GST history reconstruction from temperature profiles measured in boreholes has been extensively conducted for studies of climate change and environment change. This method is considered to be effective in investigation of long-term variation of thermal environment associated with city development, including the heat island effect. However, detailed analysis of subsurface temperature distribution in urban areas has not been made in many cities.

We have conducted measurements of subsurface temperature profiles in groundwater monitoring wells mainly in the Bangkok, Jakarta, and Taipei areas. Reconstruction of GST histories through inversion analysis with a multi-layer model was made on selected temperature profiles which are not apparently disturbed by groundwater flow. In the Bangkok area, GST history for the last 300 years was estimated at six sites located in a variety of environment ranging from the central part of Bangkok to rural agricultural land. GST has increased in the last 100–150 years at all the sites but the amount of temperature increase and the onset time of warming vary by site. The temperature increase in the central part of Bangkok amounts to about 2.5 K, while it is no more than 0.5 K at a rural site.

This result, together with the onset time of warming, well corresponds to the urbanization process of the Bangkok metropolitan area.

Reconstructed GST history in urban areas may reflect not only SAT variation due to the heat island effect but also influence of land use change associated with urbanization. It is necessary to examine land use in the past around the site where GST history was reconstructed and the relationship between land use and GST. Long-term monitoring of surface soil temperature may provide useful information on the GST and SAT relation at present. Influence of regional groundwater flow on GST reconstruction analysis is another subject to be considered.

Repeated temperature profile measurements in observation wells in the Tokyo area demonstrated increase in subsurface temperature, which represents downward propagation process of GST increase. Similar repeated measurements at intervals of several to 10 years will reveal the subsurface heat island effect in other cities as well. Temperature monitoring at 30–50 m depths enables us to observe propagation process of temperature variation in real time. Temperature records we obtained in some wells showed steady increase probably due to recent GST increase and short-period variation caused by human activity. Analysis of such long-term temperature records at multiple depths provides information on mechanism of heat transfer.

Surface warming due to the heat island effect has resulted in increase in subsurface temperature, or accumulation of heat in the subsurface of urban areas. The amount of heat stored in the subsurface at any moment in the past can be calculated based on GST history at the site estimated from subsurface temperature profiles at present. It may be considered an indicator of subsurface thermal environment in urban areas and its spatial and temporal variations over a city and in its suburbs should be useful for discussion on the relation between thermal environment and city development. In the Bangkok area, we found that heat stored in the subsurface of the city center after 1900 exceeds the double of the average over continents in the Northern Hemisphere.

Acknowledgements Measurements of borehole temperature profiles and long-term monitoring of soil/borehole water temperature were conducted by Japanese and foreign members of the research project "Human Impacts on Urban Subsurface Environment", Research Institute for Humanity and Nature (RIHN) in cooperation with the Lake Biwa Museum, Water Resources Agency, Ministry of Economic Affairs, Taiwan, National Pingtung University of Science and Technology, National Cheng Kung University, especially with assistance of K. Takahashi, M. Koizumi, and V. Harcouet.

References

Bartlett MG, Chapman DS, Harris RN (2004) Snow and the ground temperature record of climate change. J Geophys Res 109:F04008. doi:10.1029/2004JF000224
Bartlett MG, Chapman DS, Harris RN (2006) A decade of ground-air temperature tracking at Emigrant Pass Observatory, Utah. J Climate 19:3722–3731
Beltrami H, Kellman L (2003) An examination of short- and long-term air-ground temperature coupling. Glob Planet Change 38:291–303

Beltrami H, Bourlon E, Kellman L, Gonzalez-Rouco JF (2006) Spatial patterns of ground heat gain in the Northern Hemisphere. Geophys Res Lett 33:L06717. doi:10.1029/2006GL025676

Bredehoeft JD, Papadopulos IS (1965) Rates of vertical groundwater movement estimated from the earth's thermal profile. Water Resour Res 1:325–328

Cermak V (1971) Underground temperature and inferred climatic temperature of the past millennium. Palaeogeogr Palaeoclimatol Palaeoecol 10:1–19

Cermak V, Safanda J, Kresl M, Dedecek P, Bodri L (2000) Recent climate warming: surface air temperature series and geothermal evidence. Stud Geophys Geod 44:430–441

Cermak V, Safanda S, Bodri L, Yamano M, Gordeev E (2006) A comparative study of geothermal and meteorological records of climate change in Kamchatka. Stud Geophys Geod 50:675–695

Dapaah-Siakwan S, Kayane I (1995) Estimation of vertical water and heat fluxes in the semi-confined aquifers in Tokyo Metropolitan area, Japan. Hydrol Processes 9:143–160

Domenico PA, Palciauskas VV (1973) Theoretical analysis of forced convective heat transfer in regional ground-water flow. Geol Soc Am Bull 84:3803–3814

Ferguson G, Woodbury AD (2007) Urban heat island in the subsurface. Geophys Res Lett 34:L23713. doi:10.1029/2007GL032324

Gonzalez-Rouco F, von Storch H, Zorita E (2003) Deep soil temperature as proxy for surface air-temperature in a coupled model simulation of the last thousand years. Geophys Res Lett 30:2116. doi:10.1029/2003GL018264

Goto S, Yamano M (2010) Reconstruction of the 500-year ground surface temperature history of northern Awaji Island, southwest Japan, using a layered thermal property model. Phys Earth Planet In 183:435–446. doi:10.1016/j.pepi.2010.10.003

Goto S, Kim HC, Uchida Y, Okubo Y (2005) Reconstruction of the ground surface temperature history from the borehole temperature data in the southeastern part of the Republic of Korea. J Geophys Eng 2:312–319

Goto S, Yamano M, Kim HC, Uchida Y, Okubo Y (2009) Ground surface temperature history reconstructed form borehole temperature data in Awaji Island, southwest Japan for studies of human impacts on climate change in East Asia. In: Taniguchi M, Burnett WC, Fukushima Y, Haigh M, Umezawa Y (eds) From headwaters to the ocean: hydrological changes and watershed management. Taylor & Francis, London, pp 529–534

Hamamoto H, Yamano M, Goto S, Taniguchi M (2009) Estimation of the past ground surface temperature history from subsurface temperature distribution – application to the Bangkok area. Butsuri-Tansa (Geophys Explor) 62:575–584 (in Japanese with English abstract)

Harris RN, Chapman DS (1997) Borehole temperatures and a baseline for 20th-century global warming estimates. Science 275:1618–1621

Huang S (2006) 1851–2004 annual heat budget of the continental landmass. Geophys Res Lett 33:L04707. doi:10.1029/2005GL025300

Huang S, Pollack HN, Wang J-Y, Cermak V (1995) Ground surface temperature histories inverted from subsurface temperatures of two boreholes located in Panxi, SW China. J Southeast Asian Earth Sci 12:113–120

Huang S, Pollack HN, Shen P-Y (2000) Temperature trends over the past five centuries reconstructed from borehole temperatures. Nature 403:756–758

Huang S, Taniguchi M, Yamano M, Wang CH (2009) Detecting urbanization effects on surface and subsurface thermal environment – a case study of Osaka. Sci Total Environ 407:3142–3152

Jessop AM (1990) Thermal geophysics. Elsevier Science, Amsterdam, Netherlands, 306 pp

Kataoka K, Matsumoto F, Ichinose T, Taniguchi M (2009) Urban warming trends in several large Asian cities over the last 100 years. Sci Total Environ 407:3112–3119

Lachenbruch AH, Marshall BV (1986) Changing climate: geothermal evidence from permafrost in the Alaskan Arctic. Science 234:689–696

Lewis T, Wang K (1992) Influence of terrain on bedrock temperature. Palaeogeogr Palaeoclimatol Palaeoecol 98:87–100

Lewis TJ, Wang K (1998) Geothermal evidence for deforestation induced warming: implications for the climatic impact of land development. Geophys Res Lett 25:535–538

Lubis RF, Miyakoshi A, Yamano M, Taniguchi M, Sakura Y, Delinom R (2009) Reconstructions of climate change and surface warming at Jakarta using borehole temperature data. In: Taniguchi M, Burnett WC, Fukushima Y, Haigh M, Umezawa Y (eds) From headwaters to the ocean: hydrological changes and watershed management. Taylor & Francis, London, pp 541–545

Miyakoshi A, Uchida Y, Sakura Y, Hayashi T (2003) Distribution of subsurface temperature in the Kanto Plain, Japan; estimation of regional groundwater flow system and surface warming. Phys Chem Earth 28:467–475

Miyakoshi A, Hayashi T, Monyrath V, Lubis RF, Sakura Y (2009a) Subsurface thermal environment change due to artificial effects in the Tokyo metropolitan area, Japan. In: Taniguchi M, Burnett WC, Fukushima Y, Haigh M, Umezawa Y (eds) From headwaters to the ocean: hydrological changes and watershed management. Taylor & Francis, London, pp 547–552

Miyakoshi A, Hayashi T, Kawai M, Kawashima S, Hachinohe S (2009b) Change in groundwater and subsurface thermal environment in the Tokyo metropolitan area. Abstracts Japan Geophysics Union Meeting 2009, H129-P003 (abstract)

Pollack HN, Chapman DS (1993) Underground records of changing climate. Sci Am 268:16–22

Pollack HN, Huang S (2000) Climate reconstruction from subsurface temperatures. Annu Rev Earth Planet Sci 28:339–365

Pollack HN, Demezhko DY, Duchkov AD, Golovanova IV, Huang S, Shchapov VA, Smerdon JE (2003) Surface temperature trends in Russia over the past five centuries reconstructed from borehole temperatures. J Geophys Res 108(B4):2180. doi:10.1029/2002JB002154

Safanda J, Rajver D, Correia A, Dedecek P (2007) Repeated temperature logs from Czech, Slovenian and Portuguese borehole climate observatories. Clim Past 3:453–462

Smerdon JE, Pollack HN, Cermak V, Enz JW, Kresl M, Safanda J, Wehmiller JF (2004) Air-ground temperature coupling and subsurface propagation of annual temperature signals. J Geophys Res 109:D21107. doi:10.1029/2004JD005056

Smerdon JE, Pollack HN, Cermak V, Enz JW, Kresl M, Safanda J, Wehmiller JF (2006) Daily, seasonal, and annual relationships between air and subsurface temperatures. J Geophys Res 111:D07101. doi:10.1029/2004JD005578

Taniguchi M, Uemura T (2005) Effects of urbanization and groundwater flow on the subsurface temperature in Osaka, Japan. Phys Earth Planet Inter 152:305–313

Taniguchi M, Shimada J, Tanaka T, Kayane I, Sakura Y, Shimano Y, Dapaah-Siakwan S, Kawashima S (1999) Disturbances of temperature-depth profiles due to surface climate change and subsurface water flow: 1. An effect of linear increase in surface temperature caused by global warming and urbanization in the Tokyo metropolitan area, Japan. Water Resour Res 35:1507–1517

Taniguchi M, Uemura T, Jago-on K (2007) Combined effects of urbanization and global warming on subsurface temperature in four Asian cities. Vadose Zone J 6:591–596

Taniguchi M, Shimada J, Fukuda Y, Yamano M, Onodera S, Kaneko S, Yoshikoshi A (2009) Anthropogenic effects on the subsurface thermal and groundwater environments in Osaka, Japan and Bangkok, Thailand. Sci Total Environ 407:3153–3164

Uchida Y, Sakura Y, Taniguchi M (2003) Shallow subsurface thermal regimes in major plains in Japan with reference to recent surface warming. Phys Chem Earth 28:457–466

Wang K (1992) Estimation of ground surface temperatures from borehole temperature data. J Geophys Res 97:2095–2106

Wang K, Lewis TJ (1992) Geothermal evidence from Canada for a cold period before recent climatic warming. Science 256:1003–1005

Woodbury AD, Bhuiyan AKMH, Hanesiak J, Akinremi OO (2009) Observations of northern latitude ground-surface and surface-air temperatures. Geophys Res Lett 36:L07703. doi:10.1029/2009GL037400

Yamanaka T, Mikita M, Tsujimura M, Lorphensri O, Shimada J, Hagihara A, Ikawa R, Kagabu M, Nakamura T, Onodera S, Taniguchi M (2009) Assessment of enhanced recharge of confined groundwater in and around the Bangkok metropolitan area: numerical experiments and multiple tracer studies. In: Proc. international symposium on efficient groundwater resources management (IGS-TH 2009), Bangkok, Thailand, February 2009, CD-ROM

Yamano M (2010) Reconstruction of subsurface thermal environment in urban areas In: Taniguchi M (ed) Subsurface environment in Asia, Gakuho-sha, Tokyo, pp 187–213 (in Japanese)

Yamano M, Goto S, Miyakoshi A, Hamamoto H, Lubis RF, Vuthy M, Taniguchi M (2009) Reconstruction of the thermal environment evolution in urban areas from underground temperature distribution. Sci Total Environ 407:3120–3128

Chapter 12
Urban Warming and Urban Heat Islands in Taipei, Taiwan

Yingjiu Bai, Jehn-Yih Juang, and Akihiko Kondoh

Abstract This study has two main purposes; the first is to clarify urban warming in Taipei City based on 28 years of climatological data, and the second is to characterize the urban heat island (UHI) mechanism in a tropical basin using the available relevant climatological data collected from Taipei City and neighboring areas (Taipei County).

Taipei City has urbanized rapidly since 1967, and urban warming appeared from 1985. The effects of urbanization on local weather and climate change resulted in a remarkable increase in mean and minimum temperatures. However, urbanization resulted in little change in maximum temperature in Taipei City. The increase in minimum temperature in summer is significantly large in Taipei City.

The results of field observations in 2008–2009 proved that the nocturnal UHI phenomenon is predominant; however, in the rainy season (November and December), the UHI intensity during the daytime is higher than at night. The maximum UHI intensity reached 4.0–5.0°C during clear day-sky and calm wind conditions, mainly in the wet winter. In addition, during the dry months (spring), the nocturnal UHI reached its greatest intensity on cloudless nights before sunrise, and the maximum UHI intensity reached about 2.0°C.

Y. Bai (✉)
Tohoku University of Community Service and Science, 3-5-1 Iimoriyama, Sakata, Yamagata 998-8580, Japan
e-mail: bai@koeki-u.ac.jp

J.-Y. Juang
National Taiwan University, No. 1, Sec. 4, Roosevelt Road, Taipei 10617, Taiwan, ROC

A. Kondoh
Chiba University, 1-33 Yayoi-cho, Inage-ku, Chiba 263-8522, Japan

12.1 Introduction

Urbanization and urban sprawl are dominant factors in regional landscape evolution across the world and can significantly affect local climate. One of the most well-known phenomena associated with inadvertent climate change is the urban heat island (UHI), in which the air temperature in the urban canopy is higher than that in the surrounding rural area. Oke (1997) reported that the annual mean air temperature of a city with one million or more people can be 1–3°C (1.8–5.4°F) warmer than its surroundings.

Urban warming increases in intensity and area with rapid urbanization resulting from the large-scale development of commercial, manufacturing, and transportation areas. UHI intensity varies with urban size, urban surface characteristics, anthropogenic heat release, topography, and meteorological conditions (e.g., Oke 1987). Even smaller cities and towns will produce UHIs, though the effect often decreases as city size decreases (e.g., Fujino and Asaeda 1999). The relationship between UHI intensity and population is of particular scientific interest (e.g., Gyr and Rys 1995; Yoshino and Yamashita 1998). UHIs and urban warming are considered major problems faced by human beings in the twenty-first century as a result of urbanization and industrialization of human societies.

Many observational studies in mid-latitude and northern climatic zones have indicated that the UHI is prominent on calm, clear nights, and its intensity is weakest in summer and strong in fall and winter (e.g., Klysik and Fortuniak 1999; Montavez et al. 2000). However, UHI research in large cities in low-latitude regions has just begun. Several studies (e.g., Jauregui 1997) observed both nocturnal and diurnal UHI occurrences, but the nocturnal occurrence predominates. In the tropics, Jauregui (1997) reported that there is a maximum UHI of 7.8°C in Mexico City (19°26′N, 99°7′W) during clear sky and calm wind conditions, mainly in the dry winter months. It occurs diurnally during rainy months and is probably due to differences in the evaporation rate between seasons. In the subtropics, Yow and Carbone (2006) investigated the UHIs of Orlando, Florida, in the USA (28°33′N, 81°20′W). They showed that Orlando's UHIs are predominantly a nocturnal phenomenon, but with intense heat islands occurring occasionally during warm afternoons. The diurnal events are most likely attributable to isolated thundershowers. These are unexpected considering the classical descriptions of UHIs in mid-latitude cities.

Taipei City (25°05′N, 121°33′E) is located in a subtropical basin. Because of the unique landforms of the geological basin in this typhoon area, the typhoon-fed floods are enormous in these areas. In a recent study, Wang et al. (2008) documented that a strong warming trend in the Taipei basin (two times higher than the world average) was observed in the period from 1897 to 2006, which accelerated after 1980. The UHI intensity of the Taipei basin reveals an increasing trend with a monthly average of 0.011°C during 1994–2006, and during 2002–2006, the UHI anomalies show the most significant increases. However, the nocturnal and diurnal UHI phenomenon were not described in those previous studies, due to the lack of detailed record from an adequate network of observations in the city.

Issues naturally arise as to how intense and common nocturnal and diurnal UHI occurrences are compared to the corresponding description for mid-latitudes.

How does the UHI affect the number of precipitation days and intensity of precipitation?

This work presented in this chapter consists of two parts. The first part focuses on urban warming in Taipei City based on 28 years of climatological data. The second part documents the UHI mechanism in a tropical basin using the available relevant climatological data collected from Taipei City and neighboring areas (Taipei County). The main objective of this study is to determine the effect of urban development of Taipei metropolitan area on the regional climate.

12.2   Study Region and Data Description

12.2.1   Study Region

Taipei is the political, economic, and cultural center of Taiwan. The climate is affected by the East Asian monsoon and further complicated by a subtropical basin. The Danshui River, merging with the Dahan, Xindian, and Keelung Rivers, forms the Taipei basin area of 2,726 km^2. Taipei City (25°05′N, 121°33′E) and neighboring areas (Taipei County) cover an area of 380 km^2, and approximately 6.4 million inhabitants are located in the center of the Taipei basin (Fig. 12.1).

In 1932, the city covered an area of 66.98 km^2 and the population was about 600,000 people. Taipei City urbanized rapidly after 1967, and urbanization has been extensive. At present, the area of Taipei City has increased to 272.14 km^2 and the population has reached 2.62 million people (2008). Taipei County controls ten cities, four urban townships, and 15 rural townships. The population had been growing at a rapid rate in Taipei County from 1950, especially in Banciao City, Zhonghe City, Sanchong City and Yonghe City. In 2008, the population has reached 3.88 million people.

Taipei City 2.62 million peo.(2008)		Taipei County 3.84 million peo. (2008)	
District Name	Density (peo./km^2)	City Name	Population (peo.)
1 Daan	27,385	① Banqiao	553,666
2 Datong	21,718	② Zhonghe	414,849
3 Wanhua	21,240	③ Xinzhuang	401,306
4 Sonshan	22,385	④ Sanchong	389,201
5 Zhongzheng	20,860	⑤ Xindian	295,171
6 Xinyi	20,082	⑥ Tucheng	238,806
7 Zhongshan	15,945	⑦ Yonghe	236,824
8 Neihu	8,476	⑧ Luzhou	197,524
9 Wenshan	8,272	⑨ Xizhi	187,961
10 Nangang	5,171	⑩ Shulin	173,021
11 Shilin	4,547	⑪ Danshui	141,130
12 Beitou	4,360	(Township)	

Fig. 12.1 Taipei City and neighboring areas (Taipei County ①–⑪) (downtown area: districts 2, 3, 5, 7; new inner city area: districts 1, 4, 6; new urban area: districts 8–12). *Source*: Taiwan City Statistical Year Book 2009. Banciao City Household Registration Office, Taipei County. (Wade-Giles: Banciao = Pinyin: Banqiao)

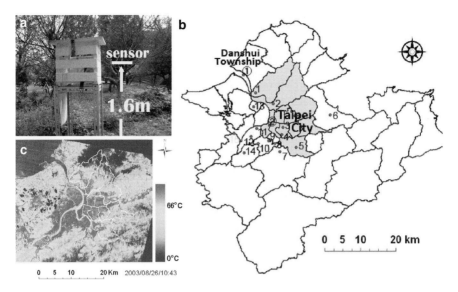

Fig. 12.2 (**a**) Thermal recorder installation. Thermometry is put in a breezy shutter. The sensor is 1.6 m above the ground. (**b**) Locations of meteorological observatories selected for this study. The distance between Danshui Township observatory (①) and Taipei City observatory (②) is about 20 km. Fifteen thermal recorder installations are established in Taipei City and Taipei County (the point of 1–15). (**c**) Satellite thermal image of Taipei metropolitan (ASTER data 2003/08/26)

12.2.2 Data Description

The data analyzed in this study includes: (1) the surface air temperature, precipitation, humidity, depression, and wind speed/direction collected at Taipei metropolitan's meteorological observatories by the Taiwan Central Weather Bureau (CWB) Services, who have the longest records; (2) temperature/humidity data collected every 10 min at the thermal recorder installations in Taipei City and Taipei County, established in 2008 by this research project (Fig. 12.2).

12.3 Warming Trend in Taipei

12.3.1 Urban and Rural Climate Data

A long-term temperature comparison between a highly urbanized area and a deurbanized area is the most common approach to analyzing urban warming. According to population growth data (1945–2009) and building data on the floor areas of newly constructed houses (Fig. 12.3), the Danshui Township Observatory (25°10′N, 121°26′E elevation = 19.0 m above mean sea level; Fig. 12.2, ①) is identified as a deurbanized site, and Taipei City Observatory (25°02′N, 121°31′E elevation = 5.3 m above mean sea-level; Fig. 12.2, ②) is identified as a highly urbanized site.

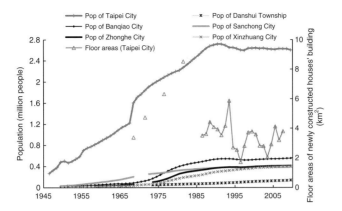

Fig. 12.3 Population growth of Taipei City and neighboring areas, and floor area increase of newly constructed houses (1945–2009). *Source*: Taiwan City Statistical Year Book 2010 and Banciao City Household Registration Office, Taipei County

Danshui is a small coastal town located to the north of the Taipei basin (Fig. 12.1). It has a population of 135 thousand people (density: 1,996 people/km², 2008). By the twentieth century, the local economy had switched from primarily fishing to agriculture. In the last decade, the town became popular in the local real estate market as a suburb of Taipei, following the completion of the Taipei Rapid Transit System's Danshui Line.

12.3.2   Warming Trend in Taipei

Several studies have shown that the warming region in Asia extends eastward to Japan and southeastward to Taiwan along the coast of China. In general, the rates of increase are larger than 1.0°C/100 years, and in higher latitudes, are larger in winter than in summer (e.g., Jones et al. 1999; Wallace et al. 1996). Figure 12.4 shows that the annual mean temperature in Taipei increased between 1897 and 2009; the warming rate is statistically 1.57°C/100 years. The rate of increase in annual mean temperature in Taipei is lower than that in Tokyo (2.45°C/100 years), a mega-city with a population of 12.3 million (2005) and Osaka (2.01°C/100 years), a mega-city with a population of 2.63 million (2005). However, the warmest monthly temperature in Taipei shows a fast rising rate, particularly during 1970–2009 (Fig. 12.4b), which is similar in the rate of Tokyo and Osaka. There is little or no increase in the coldest monthly temperature in Taipei (Fig. 12.4c), but the coldest monthly temperature in Tokyo and Osaka shows a significant increase during the same period. The results seem to be at odds with previous results (e.g., Jones et al. 1999).

The results shown here, however, indicate that the summertime warming rate can be higher at low latitudes where the mechanisms are different from the dominant mechanisms in the warming of the middle and high latitudes in the tropics.

Hsu and Chen (2002) reported that the precipitation has shown a tendency to increase in northern Taiwan and to decrease in southern Taiwan in the past 100 years.

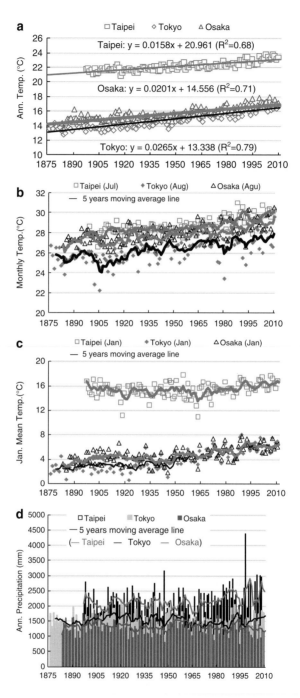

Fig. 12.4 Tendency to increase in annual mean temperature (**a**), the warmest monthly temperature (**b**), the coldest monthly temperature (**c**) and annual precipitation (**d**) in Taipei (*Source*: Taiwan Central Weather Bureau (TCWB)), Tokyo and Osaka (*Source*: Japan Meteorological Agency), (1876–2009)

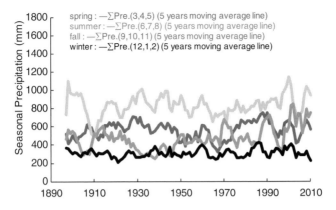

Fig. 12.5 Tendency to increase in seasonal precipitation in Taipei City (1897–2009). *Source*: Taiwan Central Weather Bureau (TCWB)

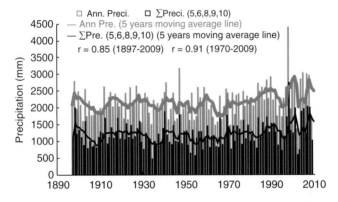

Fig. 12.6 Tendency to increase in annual mean temperature and precipitation in last spring and fall in Taipei City (1897–2009). *Source*: Taiwan Central Weather Bureau (TCWB)

Figure 12.4 indicates a marked increase in the annual precipitation in Taipei City during 1897–2009, especially after 1980. However, there is no absolute increase in annual precipitation in Tokyo and Osaka during 1875–2009. In addition, an examination of the increasing seasonal precipitation indicates that the precipitation increase in spring and in fall can be attributed to the annual precipitation increase (Figs. 12.5 and 12.6). Figure 12.6 shows a significant correlation ($r=0.85$ during 1897–2009; $r=0.91$ during 1970–2009) between annual precipitation and a total of monthly precipitation of May, June, August, September and October. Hsu and Chen (2002) pointed out that the changes in temperature and precipitation are consistent with the weakening of the East Asian monsoon.

Previous studies (e.g., Hsu and Chen 2002) documented that the temperature increase across the whole of Taiwan Island occurred most significantly during the warm season and the warming trend of the annual mean temperature is consistent with the regional warming trend pattern. In this study, the reason for this is revealed by the analysis of climatological data for 1970–2009.

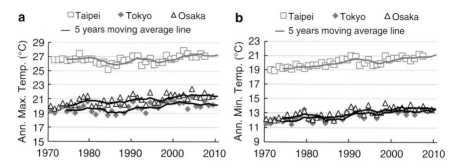

Fig. 12.7 Tendency to increase in annual maximum and minimum temperatures in Taipei (*Source*: Central Weather Bureau Ministry of Transportation and Communications), Tokyo and Osaka (1970–2009) (*Source*: Japan Meteorological Agency)

Taiwan's rapid economic growth from 1970 to 2000 transformed the island from a rural economy to an industrialized one. At the same time, there was rapid population growth in urban areas so that nearly 80% of Taiwan's people currently live in urban areas, making it an urbanized nation. Many rural areas now have croplands interspersed with small manufacturing factories or large industrial districts. At the same time, rapid population growth in urban areas has led to an increase in the number of houses being built. Figure 12.3 shows population growth in Taipei City and neighboring areas (Banqiao, Sanchong, Zhonghe, and Xinzhuang Cities, and Danshui Township, and the increase in floor area of newly constructed houses in Taipei City (1945–2009).

Figure 12.7 indicates that the increase in annual minimum temperature is larger than the increase in annual maximum temperature during 1970–2009. It is different between Taipei City and the cities in Japan. The annual minimum temperature has been increasing according to the linear regression for the last 38 years at a rate of 0.5°C/10 years. The rate of increase is little higher than both Tokyo and Osaka.

The precipitation trend during 1970–2009 is also much different from the cities in Japan (Fig. 12.4). The annual precipitation has absolutely increased in Taipei City; however, there is no absolute precipitation increase in Tokyo and Osaka. Additionally, during 1970–2009, the number of precipitation days in Taipei City has not increased. That is considered one of possible reasons that the increase in heavy precipitation in late spring and in fall is attributed to the annual precipitation increase, which is defined as a day when the daily precipitation exceeds 50 mm, has been more frequent in Taipei City since 1998. This is affected both by fluctuations in the East Asian monsoon and by local urban warming.

Overall, the increase in minimum temperature in summer is significantly large in Taipei City. Figure 12.8 indicates that the number of days with a minimum temperature ≤10°C has decreased.

Furthermore, this study compares the warming trend in temperature between a highly urbanized area and a deurbanized area using long-term data collected at the Taipei City and Danshui Township observatories (Fig. 12.9).

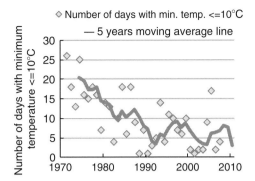

Fig. 12.8 Number of days with min. temp. $\leq 10°C$ in Taipei City (1971–2009). *Source*: Central Weather Bureau Ministry of Transportation and Communications

As of the end of 1968, there were 37,950 business firm registered in Taipei City. By the end of 2008, the number of business firms registered in Taipei City has increased to 216,758. The increase was 178,808 firms over a period of 40 years or an annual growth of 4,600 firms. The total capitalization of businesses registered in Taipei City was amounted to 13,193.360 billion NTD, an increase of 417.15 times than the 31.552 billion NTD of 1968. At the end of the 60s, the industrial factories were concentrated in the inner city. The dominant factories in the inner city were those for food products, printing & publishing. As rapid urban population growth in the industrial area (the inner city) from 1968 to 1980, the inner city and new inner area has expanded (Figs. 12.1 and 12.3). In the 80s, the industrial factories moved to the new districts of Nangang, Neihu and Shilin in the northeast and east of Taipei City (Fig. 12.1). As a result, the inhabitants of the inner city and new inner city has decreased and moved into the new urban area, so that the land-use and urban structure (town structure) have become changed in the new districts. The inner city transformed from a mixed residential area into a commercial area at the end of the 80's. Now, the manufactures of electronic equipment, chemicals and printing & publishing are the most important industries in the peripheral areas (surrounding areas) of Taipei City.

The comparison of temperature data for 1971–2009 showed that urban warming in Taipei City started in 1985. The major reason is environmental changes, such as urbanization, population growth. The effects of urbanization on local weather and climate change resulted in a remarkable increase in mean and minimum temperatures. However, urbanization resulted in little change in maximum temperature in Taipei City. Similar results were also found in case studies in Japan (e.g., Noguchi 1994).

The difference in precipitation at the Taipei and Danshui observatories is large. Figure 12.9 indicates that the mean annual precipitation in Taipei City substantially increased after 1998; however, there was no corresponding increase in the annual number of rainy days. There were notable changes in precipitation at the Danshui Township Observatory (deurbanized area). The reasons for the precipitation

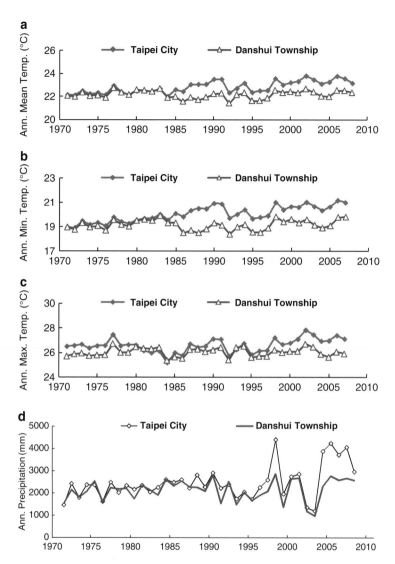

Fig. 12.9 Comparison of (**a**) ann. mean temp., (**b**) ann. min. temp., (**c**) ann. max. temp. and (**d**) ann. precipitation at Taipei City and Danshui Township Observatories (1971–2009). *Source*: Central Weather Bureau Ministry of Transportation and Communications

increase in Taipei City may be complicated by various local factors. Relatively few studies have examined the effect of urbanization on storms and precipitation. Huff and Changnon (1973) investigated nine U.S. cities and observed urban-induced increases in precipitation for all except two cities. Some studies have examined how UHIs affect urban precipitation (e.g., Jauregui and Romales 1996). It has been shown (Bornstein and Lin 2000) that UHI may initiate thunderstorms.

12.4 Urban Heat Islands in Taipei

It is well known that accurate representative real-time urban and rural climate data are the most important for UHIs studies. Furthermore, the definition of "urban temperature" is problematic. Commonly, "land-use" and DID (densely inhabited district) data are used to distinguish the urban area and the surroundings.

In this study, 15 thermal recorder installations were established in Taipei City and Taipei County according to "land-use" and DID data (Table 12.1 and Fig. 12.1). Figure 12.2 shows a thermal recorder (Hioki E.E. Corporation: **HUMIDITY LOGGER 3641-20**; temperature: ±0.5°C for 0.0–35.0°C; humidity: ±5% RH at 25°C) installation established in a green space. The thermometer measuring air temperature and humidity every 10 min is placed in a well-ventilated instrument shelter, 1.6 m above the ground. Temperature and humidity data are downloaded once every 50 days and compiled in a real-time database to characterize the Taipei UHI.

Yonghe City is most intensive area of the crowd in Taiwan (Table 12.1). Because Yonghe City covers only an area of 5.71 km², neighboring Taipei City. Three bridges connect the two cities and make Yonghe residents easily access Taipei City, so that over 80% of Yonghe city residents work in Taipei City. As Taipei City had developed rapidly during 1970–2000, Yonghe City has become the main industrial and residential district (bedroom suburbs) of Taipei City, as the other county-controlled cities (e.g. Banqiao City, Zhonghe City). However in comparison to the advancement of urbanization of Taipei City, Yonghe City, Banqiao City and Zhonghe City, where the DID is higher than other area, are not identified as highly urbanized areas.

In this study, Taking into account the urbanization economic indicators and DID, the temperature collected at Daan District (Fig. 12.2, point 4; Fig. 12.1: 1 DID: 27,385 people/km²) was selected as the "urban temperature" (Fig. 12.10: Tu), the temperature collected at Xizhi City (Fig. 12.2 point 6; Fig. 12.1: ⑨ DID: 2,639 people/km²) was selected as the "suburb temperature" (Fig. 12.10: T6).

Table 12.1 DID of Taipei city and its surrounding areas

Area name	DID (people/km²) (2008)	Area name	DID (people/km²) (2008)
Downtown area (Fig. 12.1: 2,3,5,7)	16,000–21,718	Yonghe City (Fig. 12.1: ⑦)	41,448
New inner area (Fig. 12.1: 1,4,6)	20,000–27,385	Banqiao City (Fig. 12.1: ①)	23,930
New urban area (Fig. 12.1: 8, 9,10,11,12)	4,360–8,470	Sanchong City (Fig. 12.1: ④)	23,852
Luzhou City (Fig. 12.1: ⑧)	23,738	Zhonghe City (Fig. 12.1: ②)	20,594
Xindian City (Fig. 12.1: ⑤)	2,455	Xinzhuang City (Fig. 12.1: ③)	20,331
Tucheng City (Fig. 12.1: ⑥)	8,079	Shulin City (Fig. 12.1: ⑩)	5,142
Xizhi City (Fig. 12.1: ⑨)	2,639	Danshui Township (Fig. 12.1: ⑪)	1,996

Source: Taipei City Statistical Year Book 2009
Banciao City Household Registration Office, Taipei County

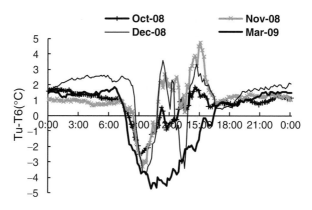

Fig. 12.10 Temperature difference between inner city (Tu) and eastern suburb (T6). (Tu: Temperature at point 4; T6: Temperature at point 6 in Fig. 12.2)

The environmental conditions affect the temperature. In order to be under the similar environmental conditions, all thermal recorder installation were established in green space in middle schools or elementary schools. However, the school's life schedule may affect the local temperature (such as club activities after school). In addition, because of the shadow effects in high-rises, the temperature at point 4 (Tu) may become lower in the morning. The temperature at point 6 (T6) may become lower in the afternoon because of the shadow effects of tall trees.

The results of field observations in 2008–2009 indicated the following: (1) the nocturnal UHI phenomenon is predominant, but in the rainy season (November and December), the UHI intensity during daytime is higher than at night; (2) the nocturnal occurrence of UHIs reached its greatest intensity on cloudless nights before sunrise, and the maximum UHI intensity reached about 2.0°C during the dry season (spring); (3) the diurnal UHI reached its greatest intensity around 12:00 and 15:00, and the maximum UHI intensity reached 4.0–5.0°C under clear day-sky and calm wind conditions, mainly in the wet winter (Fig. 12.10). This is distinctly different from what has been commonly observed in mid-latitude cities. The subtropical UHI in the Taipei basin occurs during the night only in the dry season.

Figure 12.10 indicates the diurnal UHI occurrences in Oct–Dec 2008 in Taipei City. Several studies in low-latitude regions (e.g., Jauregui 1997) observed both nocturnal and diurnal UHI occurrences, but the nocturnal occurrence predominates.

Figure 12.11 indicates that the temperature difference ($\Delta T = Tu - T11$) between the inner city (Temperature: Tu) and the satellite city (Temperature: T11) during the rainy season is small at night. However, the temperature difference during the daytime is more than 4.0°C. Downtown and its western neighboring areas (satellite cities in Taipei County) have become expanding high-temperature regions during clear day-sky and calm wind conditions during the wet winter. It imaged that the distribution of air temperature is a large circle pattern from "downtown-western neighboring areas" to the surrounding areas. Figure 12.12 is a sample of distribution

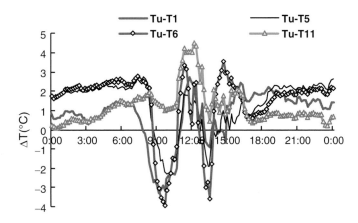

Fig. 12.11 Change of temperature difference between inner city and it surrounding areas (Dec 2008). (T1, T5, T6, and T11: Temperature at points 1, 5, 6, and 11 in Fig. 12.2)

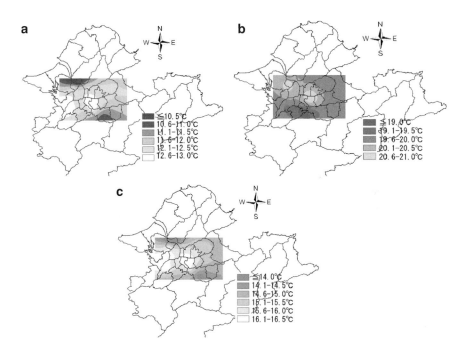

Fig. 12.12 A sample of air temperature distribution in downtown and its surrounding area (Nov 29, 2008) (**a**) 00:00 wind speed: 0 m/s (**b**) 15:00 wind speed/direction: 4.4 m/s ENE (**c**) 22:00 wind speed/direction: 1.9 m/s ESE

of air temperature during clear day-sky in November 2008. However, further detailed investigation and research on seasonal changes, daily changes with time and area, and precipitation transformation in similar meteorological conditions to the studies on mid-latitudes are needed.

Why the temperature difference between downtown and its western neighboring areas (satellite cities in Taipei County) is little during the nighttime? Why downtown and its western neighboring areas have become an expanding high-temperature region during clear day-sky and calm wind conditions (Fig. 12.12)? There are two reasons. First, those areas are the most densely populated areas in Taiwan. In just 27 years (1981–2008), the population of Taipei metropolitan area (Taipei City and neighboring area) increased 59.5%, from 4.2 to 6.7 million, which is equivalent to one-third of the total population of Taiwan (Figs. 12.1 and 12.3). Especially, population growth in Taipei County around Taipei City has been continual. Second, the region has become more urbanized in character because of its proximity to the Taipei Rapid System. At present, the rapid transit network covers Taipei City and Taipei County; therefore, the satellite cities can be considered as the central habitation area of Taipei City.

12.5   Conclusions

This study examined the increasing annual mean temperature in Taipei between 1897 and 2009. The warming rate is 1.57°C/100 years. As in other cities in low latitudes, the warmest monthly temperature in Taipei City shows a fast rising rate (1.74°C/100 years). In addition, precipitation increased in Taipei City during 1897–2009, especially after 1980.

Taipei City has urbanized rapidly since 1967, and urban warming appeared from 1985. The effects of urbanization on local weather and climate change resulted in a remarkable increase in mean and minimum temperatures. However, urbanization resulted in little change in maximum temperature in Taipei City. The increase in minimum temperature in summer is larger in Taipei City. In addition, the precipitation increase in last spring and in fall can be attributed to the annual precipitation increase. There is a significant correlation between the annual precipitation and the precipitation in last spring and fall. In particular, heavy precipitation was more frequent in Taipei City after 1998.

Yow and Carbone (2006) indicated that UHIs of Orlando, Florida, in the USA (28°33′N, 81°20′W) are predominantly a nocturnal phenomenon, but with intense heat islands occurring occasionally during warm afternoons. The diurnal events are most likely attributable to isolated thundershowers.

The results of field observations in 2008–2009 proved that the nocturnal UHI phenomenon in Taipei City is predominant; however, in the rainy season (November and December), the UHI intensity during the daytime is higher than at night. The diurnal UHI reached its greatest intensity around 12:00 and 15:00 and the maximum UHI intensity reached 4.0–5.0°C during clear day-sky and calm wind conditions, mainly in the wet winter. In addition, the nocturnal UHI reached its greatest intensity on cloudless nights before sunrise, and the maximum UHI intensity reached about 2.0°C during the dry months (spring). Downtown and its western neighboring areas (satellite cities in Taipei County) have become an expanding high-temperature region during clear day-sky and calm wind conditions. The effects of UHI on the number of precipitation days and the intensity of precipitation need further detailed investigation and research on seasonal changes, daily changes with time, and areas under meteorological conditions similar to studies in the mid-latitude.

This study is just a beginning. The results presented here are a foundation for new studies on urban warming in subtropical cities and nocturnal and diurnal UHI occurrences in a subtropical basin, in comparison with the corresponding descriptions for mid-latitude cities.

Acknowledgments The authors thank all students in the Surface-Atmosphere Interactions Laboratory, Taiwan University, for developing the thermal-recorder-observation system and downloading data. This research was partially supported by RIHN Project 2-4 Human Impacts on Urban Subsurface Environments. The work described in this report was funded by a grant for "geographical study" from FUKUTAKE Science & Culture Foundation.

The authors are very grateful to two anonymous reviewers for providing valuable comments on this paper.

References

Banciao City Household Registration Office, Taipei County. http://www.banciao.ris.tpc.gov.tw/ (in Chinese)

Bornstein R, Lin Q (2000) Urban heat islands and summertime convective thunderstorms in Atlanta: three case studies. Atmos Environ 34:507–516

Central Weather Bureau Ministry of Transportation and Communications, Republic of China. Summary report of meteorological data (1970–2009 or 1971–2009)

Fujino T, Asaeda T (1999) Characteristics of urban heat island at a small city in the lakeshore basin environment. Tenki 46:317–326 (in Japanese)

Gyr A, Rys F (eds) (1995) Diffusion and transport of pollutants in atmospheric mesoscale flow fields. Kluwer Academic, Dordrecht

Hsu HH, Chen CT (2002) Observed and projected climate change in Taiwan. Meteorol Atmos Phys 79:87–104

Huff FA, Changnon SA (1973) Precipitation modification by major urban areas. Bull Am Meteorol Soc 54:1220–1232

Japan Meteorological Agency. http://www.data.jma.go.jp/obd/stats/etrn/idex.php (Japanese)

Jauregui E (1997) Heat island development in Mexico City. Atmos Environ 31:3821–3831

Jauregui E, Romales E (1996) Urban effects on convective precipitation in Mexico City. Atmos Environ 30:3383–3389

Jones PD, New MG, Parker DE, Martin S, Rigor IG (1999) Surface air temperature and its changes over the past 150 years. Rev Geophys 37:173–199

Klysik K, Fortuniak K (1999) Temporal and spatial characteristics of the urban heat island of Lodz, Poland. Atmos Environ 33:3885–3895

Montavez JP, Rodriguez A, Jimenez JI (2000) A study of the urban heat island of Granada. Int J Climatol 20:899–911

Noguchi Y (1994) The effect of urbanization on the long-term trends of daily maximum and minimum temperatures. Tenki 41:123–135 (in Japanese)

Oke TR (1987) Boundary layer climates, 2nd edn. Methuen & Co. Ltd. (Routledge), London, UK, 435 pp

Oke TR (1997) Urban climates and global environmental change. In: Thompson RD, Perry A (eds) Applied climatology: principles & practices. Routledge, New York, NY, pp 273–287

Taipei City Government. Taipei City Statistical Year Book 2008–2010

Taiwan Central Weather Bureau (TCWB). http://www.cwb.gov.tw/ (in Chinese)

Wallace JM, Zhang Y, Bajuk L (1996) Interpretation of interdecadal trends in Northern Hemisphere surface air temperature. J Climate 9:249–259

Wang CH, Lin WZ, Peng TR, Tsai HC (2008) Temperature and hydrological variations of the urban environment in the Taipei Metropolitan area, Taiwan. Sci Total Environ 404:393–400

Yoshino M, Yamashita S (1998) Toshikankyougakujiten. Asakurashoten, Tokyo, 435 pp (in Japanese)

Yow DM, Carbone GJ (2006) The urban heat island and local temperature variations in Orlando, Florida. Southeast Geogr 46:297–321

Part V
Integrated Assessment of Subsurface Environments in Asia

Chapter 13
Long-Term Urbanization and Land Subsidence in Asian Megacities: An Indicators System Approach

Shinji Kaneko and Tomoyo Toyota

Abstract Many of the lessons concerning urban environmental problems are well documented and practiced in international environmental cooperation projects. However, most urban environmental issues analyzed in the past have concentrated exclusively on air pollution, surface water pollution and waste management in cities. With this in mind, we focus on uncovered subsurface environmental issues in cities, which is an emerging problem in developing countries in Asia. As a first step, we collected existing knowledge and information from the literature and synthesized it into a Driving Forces-Pressure-State-Impact-Response (DPSIR) framework (Jago-on et al. Sci Total Environ 407:3089–3104, 2009). Building on our previous work, the current analysis attempts to develop a stage model concerning the long-term relationships between urban development and the emerging subsurface environmental problem of land subsidence and to compare the differences and commonalities across Asian developing countries. With the help of the DPSIR framework, we select and quantify the relevant indicators for each component of the requisite framework. The results indicate that Taipei has successfully utilized its latecomer advantage and that Bangkok has benefited from its natural capacity for groundwater storage. In addition, we find that Jakarta and Manila lag behind the other cities in terms of both the recognition of the issue and the introduction of regulation to combat the problem.

13.1 Introduction

A city is a place where population is highly concentrated and so human activity is spatially intensive. This particular property also strengthens as the city develops economically in the long run. In the dynamic process of city development, various

S. Kaneko (✉)
Hiroshima University, 1-5-1 Kagamiyama, Higashi-Hiroshima, Hiroshima 739-8529, Japan
e-mail: kshinji@hiroshima-u.ac.jp

T. Toyota
Japan International Cooperation Agency, 10-5 Ichigaya Honmuracho,
Shinjuku-ku, Tokyo 162-8433, Japan

M. Taniguchi (ed.), *Groundwater and Subsurface Environments: Human Impacts in Asian Coastal Cities*, DOI 10.1007/978-4-431-53904-9_13, © Springer 2011

environmental problems take place sequentially. At the same time, the city is equipped with the capacity to cope with these problems alongside its development process. In some cases, specific environmental stresses and damage in the city tend to be more acute during certain stages of city development, and this phenomenon is typically represented by the environmental Kuznets hypothesis (Bai and Imura 2000). Using past battles against urban environmental problems in the developed world, we can draw the lesson that preventive measures are difficult but still cost effective compared with any ex-post countermeasures.

As urbanization is a global megatrend, no nation worldwide has developed effective policies to address its challenges. In particular, many developing countries in Asia face the formidable challenge of urbanization with large populations and rapid economic development. Importantly, although the urbanization rate in Asia is currently relatively low at 40.8% in 2007, it is projected to grow to 66.2% by 2050 (UN 2008). This suggests that some 1.8 billion new urban residents will be added over the next four decades to the urban population in Asia. Therefore, preventive measures against urban environmental problems should be considered to maximize the latecomers' advantage for developing Asia. In order for policymakers to undertake effective preventive measures, it is then important to recognize the long-term relationships between the urban development process and environmental problems and anticipate the occurrence of these important issues.

Fortunately, many of the lessons concerning urban environmental problems are well documented and already in practice in international environmental cooperation projects. However, the urban environmental issues analyzed in the past concentrate exclusively on air pollution, surface water pollution and waste management in cities. With this in mind, we focus on uncovered subsurface environmental issues in cities as a relatively new issue for developing Asia. As a first step, we collected existing knowledge and information from the literature and synthesized it into a Driving Forces-Pressure-State-Impact-Response (DPSIR) framework (Jago-on et al. 2009). Building on our previous work, the current study attempts to develop a stage model on the long-term relationship between urban development and the emerging subsurface environmental problem of land subsidence, and then compare the differences and commonalities across Asian developing countries. With the help of the DPSIR framework, we select and quantify the relevant indicators for each component of the requisite framework. This is first applied to the experience of Tokyo. Following this, and as a point of comparison, the exercise is repeated for five other Asian megacities, namely, Osaka, Taipei, Bangkok, Manila and Jakarta.

13.2 DPSIR Framework and Indicators

13.2.1 DPSIR Framework

The DPSIR framework is an extension of the PSR (Pressure-State-Response) framework developed by the OCED (Organisation for Economic Co-operation and Development) in 1993 as an environmental assessment tool to elucidate a holistic picture of the cause and effect relationship between human activities and environmental consequences. In 1999, the Environmental Bureau of European Union proposed the addition of two further factors, *Driving force* and *Impact*, to obtain a more comprehensive framework. Since then, the model has been applied widely to various environmental issues ranging from qualitative and heuristic approaches to quantitative and modeling approaches, including our earlier work (Jago-on et al. 2009).

13.2.2 Scope of the Study and the Selection of Indicators

Among the cities selected, Tokyo has the longest history of modern urban development beginning in the Meiji period, with land subsidence being first recognized in 1916. Given consideration of data availability, we set the timeframe for the comparative study as the approximately 100 years ranging from 1900 to 2005. For comparison, annual data for indicators for the period are constructed and analyzed to define the stages of major changes in the causes and effects of land subsidence for six megacities in Asia. Unfortunately, defining the city boundary often presents a challenge for international comparative studies of cities. Even for the same city, boundaries are usually changed as the city develops over the longer run. In this study, the current administrative boundary of the city is primarily used for measuring the indicators.

In the DPSIR framework, it is easier to first examine the impact of the issue. In the case of land subsidence, the impacts may be the physical damage to infrastructure, including roads and buildings. However, in practice it is extremely difficult to measure quantitatively the extent of damage over a long period using a single indicator. Therefore, following our previous work, we also set land subsidence itself as a proxy indicator of *Impact* and the change in the groundwater level as an indicator of *State*. As indicators for the *Pressure* from the changing groundwater level, we select groundwater abstraction and total water demand. This is because groundwater abstraction directly affects the groundwater level and total water demand is regarded as a factor linking groundwater abstraction and the indicators of *Driving force*. As factors underlying the increasing pressures on the level of groundwater, we employ population, income and industrial structure as indicators of overall city development. Finally, the regulation of groundwater abstraction is used to indicate *Response*. In total, the eight indicators shown in Table 13.1 are employed and measured.

Table 13.1 Selection of indicators using the DPSIR framework

DPSIR	Indicator
D: Driving force	Population, Income, Industrial structure (share of secondary industries)
P: Pressure	Total water demand, Groundwater abstraction
S: State	Groundwater level
I: Impact	Land subsidence
R: Response	Regulation of groundwater abstraction

13.3 Measuring Indicators

13.3.1 Methodology

In general, as annual data over more than a century is not well maintained and available, even for national data, it is extremely difficult to obtain the data for cities, especially those in developing countries. In this study, various estimations are performed to complete the mission data. When data are available and considered reliable and accurate, they are employed directly and used for reference to the other cities in the study where the corresponding data are unavailable. Therefore, in many cases, we use the data for Tokyo as a reference for the other cities.

For the indicators for *Driving forces* and *Pressure* to be measured consistently and systematically, a systems dynamics model is developed, as illustrated in Fig. 13.1. With this model, population, income, industrial structure, total water demand and groundwater abstraction are simultaneously estimated. In contrast, the groundwater level, land subsidence and the regulation of groundwater abstraction are measured separately and independently.

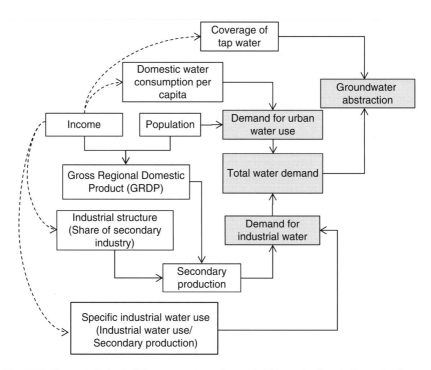

Fig. 13.1 Conceptual chart of the system dynamics model. Note: *dot lines* indicate that income is used as an explanatory variable and *gray boxes* are the various types of water volume that are estimated with the system dynamics model

13.3.2   Indicators

13.3.2.1   Population

One of the best records of socioeconomic data at the city scale over long periods is population. Even though the United Nations Population Division provides long-term datasets for the urban population of major cities across the world every 5 years, the city boundary is usually larger than the administrative boundary as it includes the population of urban agglomerations. Therefore, we collect population data for the six cities from their respective local sources. Figure 13.2 provides the results along with their source. For Tokyo and Osaka, rapid population growth began in the 1920s and continued up until the 1970s, even though they experienced a temporary drop during World War II. In contrast, the periods of rapid population growth in the other cities all began after World War II. With the exception of Manila, all cities have reached their population peak: Tokyo in the 1960s, Osaka in the 1970s, Taipei and Bangkok in the 1980s and Jakarta in the 1990s. Figure 13.2 compares the size of the cities, even though this does not necessarily represent the actual scale of the urban agglomeration.

13.3.2.2   Income

Long-term change in city wealth is one of the most important indicators from which we can infer the various changes in the socioeconomic dimensions of urban

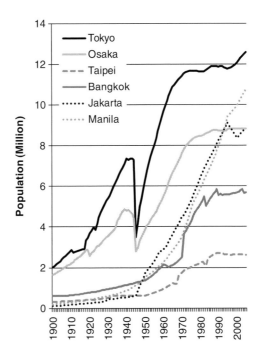

Fig. 13.2 Historical population trends. *Sources*: BPS (various) TMG (various), OPG (various), Ho (1970), TCG (2005, 2008), Larry (1982), Christopher (2007), Doeppers (1984), Sternstein (1982), NSO, UN Population Division. Note: missing data are estimated by interpolation

development. As a measure of the wealth of the city, we employ Gross Regional Domestic Product (GRDP) per capita as an indicator of the level of income for urban residents. Although some cities in our study have published official statistical data on GRDP, most of the data are relatively recent and barely cover the study period. However, long-term measures of Gross Domestic Product (GDP) are estimated by Maddison (2009) and are available for all of the countries included in the study. This provides a sound and complete reference of the level of city income and its changes over the nearly 100 years. Therefore, we begin with this national data as a benchmark to construct the data on GRDP per capita for the cities. In order to convert national income to city income, we estimate the gap between the national and city levels, as is shown in Fig. 13.3.

The two Japanese cities, Tokyo and Osaka, have relatively long-term records on GRDP from local sources, although complete data are not available. Therefore, the relationships between the gap and GDP per capita for Japan are first derived for Tokyo and Osaka. The older data for GRDP per capita in Tokyo and Osaka are then estimated using their respective relationships to construct the complete historical data. For Bangkok and Jakarta, GRDP data are also partly available from local sources. The gaps between income at the national and city levels are relatively higher for both of these cities and they jointly form an inverted U-shape with per capita national GDP as is shown in Fig. 13.3. Using this relationship, complete data

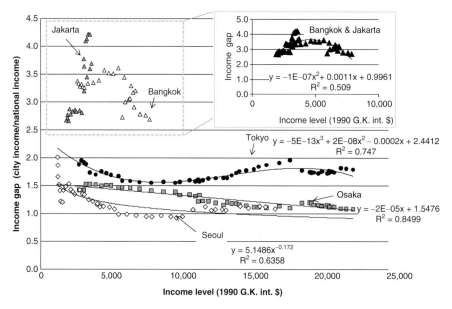

Fig. 13.3 National and city income gap. *Source*: Angus (2009), ESRI, NESDB, Pierre (2005), BPS (various), Note: Tokyo 1955–2006, Osaka 1955–2006, Bangkok 1981–2005, Jakarta 1983–2005

Fig. 13.4 Trends in income level

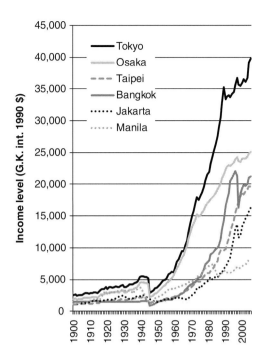

are estimated throughout the study period for Bangkok and Jakarta. For Taipei and Manila, no reliable and consistent data for GRDP are available. Accordingly, we assume that Manila displays the same relationship for the income gap that we derived from Bangkok and Jakarta; for Taipei, we estimate the corresponding relationship for Korea and Seoul as their data are readily available. We consider the similarity of Seoul and Taipei from two perspectives. The first is the primacy of Seoul and Taipei as the capital cities of their respective countries, and the second is their similar historical development process and pattern as members of the NIES (New Industrial Economies).

Figure 13.4 depicts the results of the estimation and construction of per capita GRDP for the six megacities included in this study. It should be noted that all the data in Fig. 13.4 are converted to real monetary terms as 1990 Geary–Khamis (G.K) international (int.) USD ($) as in Angus (2009).

13.3.2.3 Industrial Structure

Petty–Clark's law argues that, with economic development, macroindustrial structure shifts its relative weight from primary to secondary industry and eventually further to tertiary industry. Although this argument is usually applied to national economic structure, we assume this to be also fundamentally applicable to megacities. We should note, however, that the industrial transformation process at a certain point in economic development (as measured by income) may not be same nationally as in

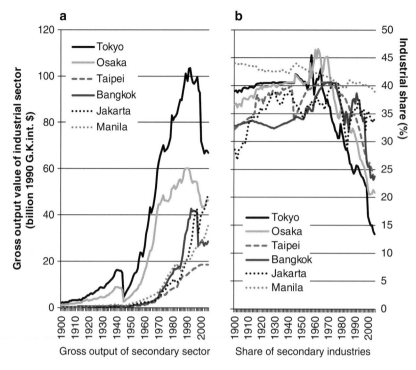

Fig. 13.5 Trends in output and the share of secondary sector. (**a**) Gross output of secondary sector. (**b**) Share of secondary industries

the city. Consequently, by using the partially available data from Tokyo, Osaka, Bangkok and Jakarta, the functional relationships between per capita GRDP and the share of secondary industry is derived and used to estimate the missing data for the share of secondary industry for the six cities from 1900 to 2005.

By incorporating the abovementioned functional relation between income and the share of secondary industry into the system dynamics model, the annual estimations of gross output value of secondary industry and its share relative to total GRDP is derived, as shown in Fig. 13.5a, b. As shown, Tokyo, Osaka and Bangkok have peaked in terms of the absolute magnitude of gross output of secondary industry, whereas the output of the other cities continues to increase. However, after the 1960s, the share of secondary industry in most of the cities started to decline, although at different speeds.

13.3.2.4   Total Water Demand

As an indicator of *Pressure*, total water demand is estimated for the six cities. Total water demand in this study is defined as the sum of urban and industrial water demand. In turn, urban water demand includes both residential and commercial

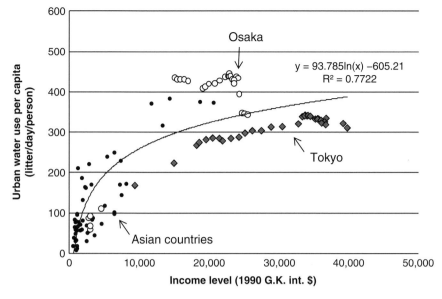

Fig. 13.6 Changing urban water use per capita. *Source*: FAO, TMG (various), OPG (various). Note: Tokyo 1965–2005, Osaka 1919–1922, 1940, 1975–2005, Asian countries (Bangladesh, Cambodia, China, Democratic People's Republic of Korea, India, Indonesia, Japan, Lao People's Democratic Republic, Malaysia, Mongolia, Nepal, Philippines, Republic of Korea, Singapore, Sri Lanka, Thailand, Viet Nam) 1980, 1985, 1990, 1995, 2000, 2005

water demand. Historically, per capita urban water demand increases with GRDP per capita in most cities, but at a diminishing rate. This logarithmic functional relation is derived from the available data for Tokyo and Osaka together with the national data for the other Asian countries (Fig. 13.6). Given this relation and total population in each city for each year, we estimate the annual urban water demand.

In contrast, industrial water demand is estimated by multiplying industrial water demand per unit of production and the total value of secondary industry output, as estimated in the previous section. Along with economic development represented by the increase in per capita income, production technology in the industrial sector has also generally progressed. This can be easily observed in the form of water-saving technology and improving water-use efficiency in the industrial sector. To address the data constraints needed to analyze this relation, we assume that the experience of Tokyo and Osaka is applicable to the other Asian megacities. We then derive a functional relationship between per capita GRDP and industrial water demand per unit of production using the data from Tokyo and Osaka, as shown in Fig. 13.7.

Figure 13.8 compares the historically estimated data for total water demand for the six megacities. As shown, total water demand peaked in Tokyo and Osaka in the 1960s, after which demand has either remain unchanged or even declined slightly. In contrast, Bangkok and Taipei reached their peak in total water demand in the late 1980s. Table 13.2 summarizes and compares the changes in the shares of industrial to total water demand.

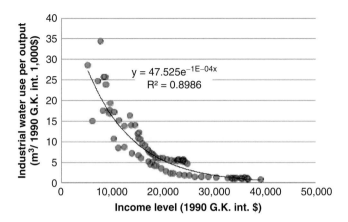

Fig. 13.7 Industrial water demands per unit of production. *Source*: METI (various). Note: Tokyo 1965–2005, Osaka 1919–1922

Fig. 13.8 Estimated total water demand

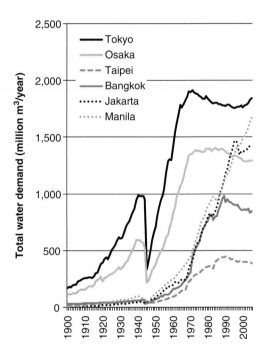

Table 13.2 Percentage share of industrial water demand (industrial water demand/total water demand in the city)

	1900	1920	1940	1960	1970	1980	1990	2000	2005
Tokyo	43.5	44.7	45.2	44.8	34.8	23.2	9.3	5.3	3.2
Osaka	42.1	43.8	45.2	48.5	41.0	31.3	22.4	15.0	13.8
Taipei	40.4	42.6	44.0	44.4	45.1	42.0	33.8	14.2	10.6
Bangkok	40.6	40.7	40.9	42.9	45.2	41.4	33.1	22.3	19.2
Jakarta	39.4	41.7	43.4	42.4	43.5	41.8	41.8	36.1	31.5
Manila	51.4	45.9	45.9	45.9	45.8	44.3	44.9	44.3	43.1

13.3.2.5   Groundwater Abstraction

Consistent and reliable long-term data for groundwater abstraction are not available. Therefore, the following method is used for the missing data. First, we collect data for the diffusion rate of tap water systems in terms of quantity of water supply, defined as the ratio of the amount of water supplied by the tap water system to total water demand. In general, the diffusion rate of the tap water system is increasing with economic development to 100% and hence this functional relation is derived from the available data in Tokyo, Osaka, Taipei and Japan, as shown in Fig. 13.9. In this study, the amount of water supplied or obtained from sources other than the tap water supply system is assumed to be groundwater abstraction. Although data on groundwater abstraction is available in some of the cities, it often covers only large users of groundwater. Therefore, there is a tendency to underestimate the data on groundwater abstraction. In such cases, we use our estimates.

Figure 13.10 provides the results of the estimation for groundwater abstraction, which is estimated by deducting the amount of piped water supply from total water demand. With the exception of Manila, groundwater abstraction in all of the study cities peaked before 2000; water abstraction peaked in Osaka even before World War II. Manila is currently estimated to abstract groundwater by more than six million m³ per year; this is larger than the peak amount in Tokyo in the 1950s.

13.3.2.6   Groundwater Level

Unlike some of the other indicators, the groundwater level cannot be estimated using functional relations with income and other socioeconomic variables as it

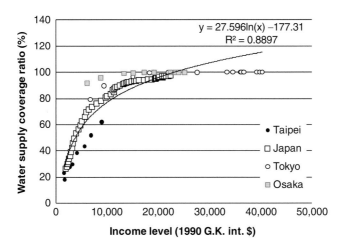

Fig. 13.9 Coverage ratios of water supply and income. *Source*: MIC and DBAS (2009). Note1: Tokyo 1965–2006, Osaka 1906–2005, Taipei 1911, 1945–1987, Japan 1905–2007. Note2: With the use of regression curve, when water supply coverage ratio goes beyond 90%, the ratio is censored at 90% as maximum since Tokyo and Osaka still utilize groundwater at least 10% of total water supply

Fig. 13.10 Estimation of
groundwater abstraction

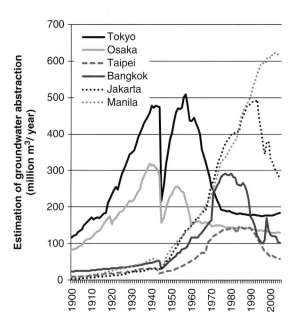

depends largely on natural, geographical and hydrological factors. In addition, the groundwater level in a city has three theoretical dimensions, namely, spatial distribution with different depths. One representation of the concept of groundwater level then requires a certain aggregation of three-dimensional data. In this study, an average of records from monitoring wells is calculated annually in each city, and we regard this as an indicator of the groundwater level.

Figure 13.11 synthesizes the available data for five of the six study cities (no data for Manila are available). First, because of the rapid increase in groundwater abstraction, the groundwater level started to decrease. Then, after groundwater abstraction was regulated, the groundwater level bottomed until it started to recover, with the various time lags depending on the conditions. The lowest groundwater levels were recorded in different periods in each city: in the middle of the 1960s in Tokyo, in the early 1970s in Osaka and in the late 1990s in Bangkok. For Taipei, although there is not sufficient data available to make a definitive judgment on the bottoming out of groundwater level, we expect this took place sometime in the late 1970s or early 1980s. The groundwater level in Jakarta continues to decline at present.

13.3.2.7 Land Subsidence

The extent of land subsidence is measured by the cumulative amount of land subsidence. Like the indicator of the level of groundwater, in nature the concept of land subsidence has three dimensions. Fortunately, continuous monitoring records of land subsidence at several points for each city are available. Therefore, we simply take the average of the annual records for different places in each city. These are synthesized in Fig. 13.12. It should be noted that comparison across different cities make little

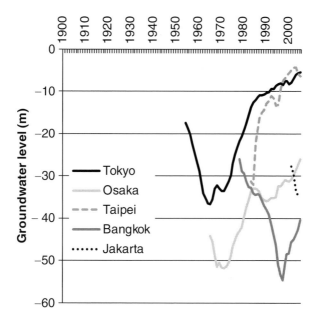

Fig. 13.11 Averaged records of groundwater level. *Source*: CEC (2009), RIEAF (various), DMR, Mining Service Jakarta Metropolitan

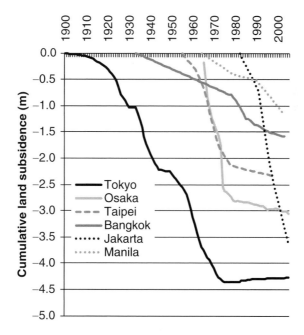

Fig. 13.12 Cumulative land subsidences (m). *Source*: Abidin et al. (2004), CEC (2009), Colbran (2009), RIEAF (various), WRA, Phien-wej et al. (2006), Piancharoen (1997), Abidin et al. (2001), Murdohardono and Sudarsono (1998), Rodolfo and Siringan (2006)

sense as geographical conditions, including soil deformability and consolidation, the number of monitoring points and the monitoring methods vary from city to city. Therefore, only the historical changes in land subsidence in each city used to observe the pace of land subsidence and the time of resolution make sense.

13.3.2.8   Regulation of Groundwater Abstraction

For measuring the indicator of *Response*, stringency and effectiveness of the regulation of groundwater abstraction, a categorical rating system is defined: (1) unawareness, (2) recognizing, (3) regulating and (4) being regulated and controlled. Based on the literature for each city, the years for the shifting of the categories are identified.

13.4   Stage Model of Urban Development and Land Subsidence

13.4.1   Tokyo Reference Model

As a reference for the other cities, this section attempts to define the stage model based on the experience of Tokyo regarding the relationship between urban development and land subsidence.

Figure 13.13 compiles the DPSIR indicators explained above over the study period from 1900 to 2005, where three factors are summarized as the dominant sector of water demand, dominant sources of water supply and the responses categories.

The first stage is defined as the period from 1900 to 1916 when land subsidence was first recognized. In this period, and as the water supply system was not well established, dependence on groundwater was high. The government-led modern industrialization progressed fully in selected strategic areas and Tokyo rapidly urbanized alongside this process as one of the key areas. At the same time, military industrial bases were formed in coastal areas of Tokyo and this further accelerated the pace of urbanization. Therefore, in some factory locations, intensive groundwater abstractions began and consequently led to an increasing incidence of land subsidence.

The second stage is then defined from the recognition of land subsidence to the introduction of effective measures against it in 1961 when land subsidence in Tokyo became seriously aggravated. The second stage in Tokyo can be further divided into two periods, namely, before and after World War II. Even before the war, when land subsidence was partly recognized by the government, it failed to attract political interest and was not given priority among the other emerging issues in the lead up to the war. Furthermore, rapid economic recovery was taking place in Tokyo through the promotion of heavy chemical industrialization, and this seriously affected the groundwater aquifer in Tokyo. During the economic recovery in Tokyo, the population increased greatly and, consequently, residential water demand also increased. Fifteen years after the end of World War II, Tokyo introduced regulations

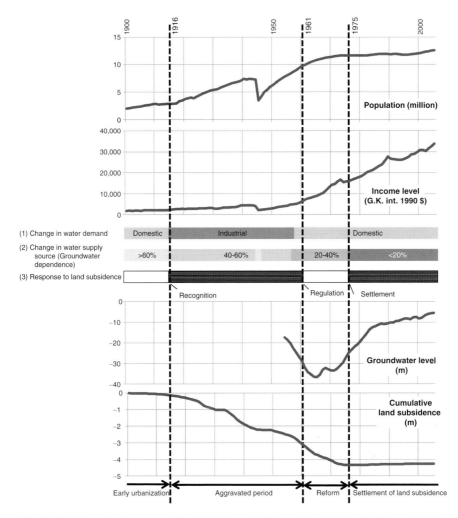

Fig. 13.13 Urban development and land subsidence in Tokyo. Note: During the War period, due to physical and operational difficulties, the piped water supply was declined and dependency on groundwater was temporarily increased at more than 60%

on groundwater abstraction when land subsidence caused serious and apparent damages to buildings and infrastructure. However, in terms of the cause and effect relation between urbanization and land subsidence, the fundamental mechanism and phenomena is not different in the pre- and post-war periods. Therefore, we combined the two periods into the one stage.

The third stage began in 1961 when the Industrial Water Law was enacted and ended in 1975 when land subsidence was almost controlled. During the third stage, industrial water works began to supply industrial water as an alternative water source, and industry in Tokyo achieved water saving through the development of water-saving technologies and the promotion of recycled water use. Moreover, economic development in Tokyo took place alongside industrial transformation when water-intensive industries progressively moved out of Tokyo. As a consequence, the dependence on groundwater in Tokyo largely declined during this period.

The fourth and final stage continues up until the present where land subsidence has been almost resolved. Nevertheless, new issues have arisen in this period. In the second and third stages, a number of facilities and infrastructures were constructed when the level of groundwater was being lowered. As a result of the regulation, when the groundwater recovered, these underground structures developed enormous upwardly buoyant forces. This has imposed a substantial financial burden to fix those structures anchored to deep foundations. Within the capacity for the natural recharge of groundwater in Tokyo, sustainable groundwater use has also emerged as a new policy issue in this stage, including the use of groundwater for watering as a countermeasure to the formation of heat islands.

13.4.2 Application to Asian Megacities

In this section, the stage model is applied to Osaka, Taipei, Bangkok, Jakarta and Manila. As industrialization in Tokyo started in 1900, we additionally defined as the zeroth stage for the pre-period of modern industrialization. The results are summarized in Table 13.3.

As shown, Osaka and Taipei have already entered the fourth stage where land subsidence is almost controlled. Bangkok on the other hand, is presently in the third stage where land subsidence is in the process of being controlled by supplying alternative water sources to industry together with relatively large capacity of groundwater storage (Taniguchi et al. 2009). Jakarta and Manila are both in the second stage where, despite the recognition of land subsidence by their respective governments, effective countermeasures have not yet been institutionalized. In Jakarta, although more than 60% of the city area is below sea level and seawater flooding takes place, a data collection system for groundwater abstraction and the regular monitoring of land subsidence is not yet fully established by the government. One of the reasons for this is the limited ability of governments to fund the provision of alternative water sources for surface water through industrial water works.

Table 13.3 Defined stages for Asian megacities

	0th stage	1st stage	2nd stage	3rd stage	4th stage
Tokyo		1900–1915	1916–1960	1961–1974	1975–
Osaka	1900–1917	1917–1928	1929–1958	1959–1979	1980–
Taipei	1900–1915	1916–1955	1956–1978	1979–1985	1986–
Bangkok	1900–1950	1951–1980	1981–1996	1997–	
Jakarta	1900–1975	1976–1992	1993[a]–		
Manila	1900–1950	1951–2001	2002–		

[a]Note on recognition: Jakarta was assessed by a Dutch surveyor in 1926, researchers in 1978, and recognized by the Mining Agency of Jakarta City in 1993

Table 13.4 Comparison of indicators for Stage 2 and Stage 3

DPSIR Indicators	Variable	Unit	Tokyo	Osaka	Taipei	Bangkok	Jakarta	Manila
Beginning of Stage 2								
Year of recognition		Year	1916	1929	1956	1965	1993	2002
Driving force: Size of population	Total urban population	Million persons	2.88	3.48	0.80	2.12	8.77	10.27
		Tokyo = 1.0	1.0	1.2	0.3	0.7	3.0	3.6
Driving force: Change in population	Change in total urban population (before and after 3 years)	%	3.5%	2.8%	4.7%	2.2%	1.4%	1.5%
		Tokyo = 1.0	1.0	0.8	1.3	0.6	0.4	0.4
Driving force: Size of economy	GRDP	1990 G.K. in billion $	9.9	11.3	1.6	6.3	73.8	75.5
		Tokyo = 1.0	1.0	1.2	0.2	0.6	7.5	7.7
Driving force: Change in economy	Change in GRDP	%	7.7%	3.6%	8.3%	10.1%	11.7%	5.1%
		Tokyo = 1.0	1.0	0.5	1.1	1.3	1.5	0.7
Driving force: Level of income	Per capita income	1990 G.K. int. $	3,416	3,260	2,034	2,992	8,415	7,354
		Tokyo = 1.0	1.0	1.0	0.6	0.9	2.5	2.2
Driving force: Change in income	Change in per capita income (before and after 3 years)	%	4.0%	0.8%	3.5%	7.8%	10.2%	3.5%
		Tokyo = 1.0	1.0	0.2	0.9	1.9	2.5	0.9
Pressure: Level of water pressure 1	Total water demand	Million cubic meter	300	351	71	202	1,342	1,543
		Tokyo = 1.0	1.0	1.2	0.2	0.7	4.5	5.1
Pressure: Change in water pressure 1	Change in total water demand (before and after 3 years)	%	6.3%	3.3%	7.6%	7.7%	3.7%	2.5%
		Tokyo	1.0	0.5	12	12	0.6	0.4
Pressure: Level of water pressure 2	Per capita water demand	Cubic meter	104.0	100.8	89.3	95.0	153.0	150.2
		Tokyo = 1.0	1.0	1.0	0.9	0.9	1.5	1.4
Pressure: Change in water pressure 2	Change in per capita water demand (before and after 3 years)	%	2.6%	3.9%	0.9%	−1.3%		
		Tokyo = 1.0	1.0	1.5	0.4	−0.5		
Pressure: Level of water pressure 3	Groundwater abstraction	Million cubic meter	182	217	41	129	493	622
		Tokyo = 1.0	1.0	1.2	0.2	0.7	2.7	3.4
Pressure: Change in water pressure 3	Change in groundwater abstraction (before and after 3 years)	%	4.5%	3.0%	6.8%	4.3%	−42%	0.1%
		Tokyo = 1.0	1.0	0.7	1.5	1.0	−0.9	0.0
Impact: Land subsidence	Land subsidence	cm	−18.2	−50.5	−7.0	−57.0	−128.0	−111.5
		Tokyo = 1.0	1.0	2.8	0.4	3.1	7.0	6.1

(continued)

Table 13.4 (continued)

DPSIR Indicators	Variable	Unit	Tokyo	Osaka	Taipei	Bangkok	Jakarta	Manila
Impact: Change in land subsidence	Change in subsidence (before and after 3 years)	cm/year Tokyo = 1.0	−2.66 1.0	−7.04 2.6	−3.42 1.3	−1.78 0.7	−24.67 9.3	−5.11 1.9
Years b/w stage 2 and stage 3			**45**	**30**	**23**	**32**	–	–
Beginning of Stage 3 Year of regulation		Year	1961	1959	1979	1997	–	–
Driving force: Size of population	Total urban population	Million persons Tokyo = 1.0	9.94 1.0	5.29 0.5	2.20 0.2	5.60 0.6		
Driving force: Change in population	Change in total urban population (before and after 3 years)	% Tokyo = 1.0	2.8% 1.0	3.8% 1.4	0.7% 0.3	0.3% 0.1		
Driving force: Size of economy	GRDP	1990 G.K. int. billion $ Tokyo = 1.0	77.1 1.0	29.3 0.4	13.2 0.2	119.3 1.5		
Driving force: Change in economy	Change in GRDP	% Tokyo = 1.0	10.3% 1.0	11.8% 1.1	4.8% 0.5	−1.0% −0.1		
Driving force: Level of income	Per capita income	1990 G.K. int. $ Tokyo = 1.0	7,757 1.0	5,537 0.7	6,028 0.8	21,289 2.7		
Driving force: Change in income	Change in per capita income (before and after 3 years)	% Tokyo = 1.0	7.3% 1.0	7.7% 1.1	4.0% 0.6	−1.3% −0.2		
Pressure: Level of water pressure 1	Total water demand	Million cubic meter Tokyo = 1.0	1,546 1.0	748 0.5	349 0.2	906 0.6		
Pressure: Change in water pressure 1	Change in total water demand (before and after 3 years)	% Tokyo = 1.0	5.4% 1.0	7.7% 1.4	1.6% 0.3	−1.0% −0.2		
Pressure: Level of water pressure 2	Per capita water demand	Cubic meter Tokyo = 1.0	155.6 1.0	141.4 0.9	159.1 1.0	161.7 1.0		
Pressure: Change in water pressure 2	Change in per capita water demand (before and after 3 years)	% Tokyo = 1.0	2.6% 1.0	3.9% 1.5	0.9% 0.4	−1.3% −0.5		

		Units				
Pressure: Level of water pressure 3	Groundwater abstraction	Million cubic meter	444	180	140	110
		Tokyo = 1.0	1.0	0.4	0.3	0.2
Pressure: Change in water pressure 3	Change in groundwater abstraction (before and after 3 years)	%	−2.9%	−6.6%	0.1%	1.5%
		Tokyo = 1.0	1.0	2.3	0.0	−0.5
Impact: Land subsidence	Land subsidence	cm	−3.3	−2.0	−2.2	−1.5
		Tokyo = 1.0	1.0	0.6	0.7	0.5
Impact: Change in land subsidence	Change in land subsidence (before and after 3 years)	cm/year	−0.05	−0.20	−0.01	−0.00
		Tokyo = 1.0	1.0	3.5	0.2	0.1

Source: Data compiled by authors from various sources. Detailed information is given in Figs. 13.2, 13.4, 13.8, 13.10, and 13.12.

13.4.3   Comparison of the Stages with Indicators

As a summary, we highlight two important milestones for the stage model in order to characterize and compare the features of each city. The first is when each city recognized land subsidence and the second is when each city introduced counter-measures against land subsidence. Figure 13.13 compares the key indicators of the DPSIR model for each city, normalized to that of Tokyo as a reference. As shown, Taipei has both recognized and taken policy measures earlier than Tokyo in terms of urban development and the severity of land subsidence. Consequently, the speed of recovery is quicker than Tokyo. Therefore, one can say that Taipei could utilize its latecomer advantage.

On the other hand, Bangkok has taken effective actions later than Tokyo, although Bangkok, like Taipei, has recognized the issue earlier than Tokyo. However, even though regulatory measures have been taken later in Bangkok, the recovery of land subsidence is quicker than that in Tokyo and Taipei because of a more favorable natural capacity. Finally, Jakarta and Manila have not recognized the issue until both cities have already developed in terms of population size and income level. Furthermore, although both these cities have reached a level of urban development comparable with the other cities, effective countermeasures have not yet taken place. One might argue for Manila that natural capacity as recharge and/or storage of groundwater is relatively large that makes effective measurements by the government delayed since the land subsidence is less revere than the other cities like Tokyo and Osaka.

13.5   Summary and Conclusions

The paper attempts to compare the long-term causal relationship between urban development and land subsidence for selected Asian megacities at different development stages and with different processes over the last century. The selection and construction of the indicators depicting the causality are conducted using the DPSIR framework developed in Jago-on et al. (2009). Concurrently, a stage model for land subsidence with urbanization is also developed. As a result of the indicators of DPSIR framework and the stage model, we can characterize each of the cities as follows: Taipei as an effective user of its latecomer advantage, Bangkok as a beneficiary of its natural capacity, and Jakarta and Manila at risk of unaddressed overdevelopment. Moreover, Osaka is characterized as a city overcoming its disadvantage in natural capacity when similar experiences in Tokyo are observed.

To cope effectively with land subsidence in developing megacities, a strategic and long-term perspective is required. A policy mix of both immediate countermeasures, such as regulation, and strategic long-term measures, such as the development of alternative water supplies, is strongly suggested for megacities when land subsidence is recognized. The timing of the various countermeasures, however, needs to be properly investigated in order to minimize any long-term social costs.

In this regard, this analysis will be of some assistance in providing policy relevant information from a comparative perspective to maximize the advantage for late-comers to this key problem.

One of the main limitations of this study is information availability. Importantly, the accumulation of more reliable and accurate information would improve the quality of analysis and consequently help to provide more concrete and specific policy prescriptions for each city. Therefore, especially for latecomers like Jakarta and Manila, advances in monitoring, scientific research and the collection of the relevant socioeconomic indicators are highly recommended.

References

Abidin H, Djaja R et al (2001) Land subsidence of Jakarta (Indonesia) and its geodetic monitoring system. Nat Hazards 28:365–387

Abidin H, Andreas H et al (2004) Land subsidence in the urban areas of Indonesia. 3rd FIG Regional Conference for Asia and the Pacific

Angus Maddison. Historical Statistics for the World Economy: 1-2006 AD. http://www.ggdc.net/MADDISON/oriindex.htm. Accessed 10 November 2010

Bai X, Imura H (2000) A comparative study of urban environment in East Asia: stage model of urban environmental evolution. Int Rev Glob Environ Strateg 1(1):135–158

BPS (Badan Pusat Statistik) (various) Statistical Yearbook of Indonesia. Jakarta: BPS-Statistics Indonesia

BPS (Statistics DKI Jakarta Provincial Office) (various) Jakarta in Figures. BSP

CEC (Civil Engineering Center) (2009) Land subsidence survey report; 2009 (Heisei Nijunen Jibanchinka Chosa Houkokusyo). Tokyo Metropolitan Government

Christopher S (2007) Planning the Megacity: Jakarta in the twentieth century. Routledge, London and New York

Colbran N (2009) Will Jakarta be the next Atlantis? Excessive groundwater use resulting from a failing piped water network. Law Environ Dev J 5(1):18–38. http://www.lead-journal.org/content/09018.pdf. Accessed 4 July 2010

DBAS (Department of Budget, Accounting & Statistics) (2009) Taipei City statistical yearbook 2008. Department of Budget, Accounting & Statistics, Taipei City Government

DMR (Department of Mineral Resources) (1992) Records of groundwater monitoring wells in Bangkok and adjacent provinces, mitigation of groundwater crisis and land subsidence project (MGL Project) report no.1. Thailand, Ministry of Industry and Public Works

Doeppers DF (1984) Manila, 1900–1941: Social Change in Late Colonial Metropolis. New Haven, Connecticut: Yale University Southeast Asia Studies.

ESRI (Economic and Social Research Institute) Cabinet Office. Annual report on national accounts. http://www.esri.cao.go.jp/en/sna/data.html. Accessed 4 July 2010

FAO (Food and Agricultural Organization of the United Nation), Land and Water Division. AQUASTAT. http://www.fao.org/nr/water/aquastat/main/index.stm. Accessed 4 July 2010

Ho SPS (1970) Economic development of Taiwan 1860–1970. Yale University Press, New Haven and London

Jago-on KAB, Kaneko S et al (2009) Urbanization and subsurface environmental issues: an attempt at DPSIR model application in Asian cities. Sci Total Environ 407:3089–3104

Larry S (1982) Portrait of Bangkok, published to commemorate the bicentennial of the capital of Thailand. Bangkok Metropolitan Administration

Maddison A (2009) Statistics on world population, GDP and Per Capita GDP, 1-2008 AD. http://www.ggdc.net/maddison/. Accessed 4 July 2010

METI (Ministry of Economy, Trade and Industry) (various) Industrial statistics survey result report. http://www.meti.go.jp/statistics/tyo/kougyo/archives/index.html. Accessed 4 July 2010

MIC (Ministry of Internal Affairs and Communications) Historical Statistics of Japan. Statistics bureau. http://www.stat.go.jp/english/data/chouki/index.htm. Accessed 4 July 2010

Murdohardono D, Sudarsono U (1998) Land subsidence monitoring system in Jakarta. In: Proceedings of symposium on Japan–Indonesia IDNDR project; volcanology, tectonics, flood and sediment hazards, Bandung, pp 243–256

NESDB (Office of the National Economic and Social Development Board). BKK & VICINITIES. http://www.nesdb.go.th/. Accessed 4 July 2010

NSO (National Statistical Office) Philippine in figure. Republic of the Philippines. http://www.census.gov.ph. Accessed 4 July 2010

OPG (Osaka Prefecture Government) (various). Osaka statistic yearbook, Osaka Prefecture

Phien-wej N, Giao PH, Nutalaya P (2006) Land subsidence in Bangkok, Thailand. Eng Geol 82:187–201

Piancharoen P (1997) Groundwater and land subsidence in Bangkok, Thailand. IAHS 121:355–364

Pierre E (2005) Indonesia's new national accounts. Bull Indones Econ Stud 41(2):253–262

Rodolfo KS, Siringan F (2006) Global sea-level rise is recognized, but flooding from anthropogenic land subsidence is ignored around northern Manila Bay, Philippines. Disasters 30(1):118–139

RIEAF (Research Institute of Environment, Agriculture and Fisheries) (various). Osaka Prefecture Environmental White Paper, Osaka prefecture

Sternstein L (1982) Portrait of Bangkok. Bangkok Metropolitan Administration

Taniguchi M, Burnett WC, Ness G (2009) Integrated research on subsurface environments in Asian urban areas. Sci Total Environ 407:3076–3088

TCG (Taipei City Government) (2005) Taipei city statistical yearbook 2004. Department of Budget, Accounting and Statistics (DBAS)

TCG (Taipei City Government) (2008) Taipei city statistical yearbook 2007. DBAS

TMG (Tokyo Metropolitan Government) (various) Tokyo statistical yearbook. Tokyo Prefecture

UN (United Nations), Population Division, Department of Economic and Social Affairs (2008) World urbanization prospects: the 2007 revision. New York

WRA (Water Resource Agency) Statistics of Water Resources. Ministry of Economic Affairs, Taiwan. http://eng.wra.gov.tw. Accessed 4 July 2010

Chapter 14
Sinking Cities and Governmental Action: Institutional Responses to Land Subsidence in Osaka and Bangkok

Takahiro Endo

Abstract The purpose of this paper is to compare land subsidence policies in Osaka and Bangkok and provide some lessons from these two cases. Land subsidence can be regarded as an example of a social dilemma. Prevention of land subsidence is a common benefit for people who suffer flood or high tide risk. However, a group of people that share such a common benefit often fail to realize it voluntarily because each individual has an incentive to be a free-rider. Governmental intervention is an effective way to address this situation, because the government can force groundwater users to change their behavior. Indeed, this was why government intervention had taken place in Osaka and Bangkok. Various policies were implemented, including designation of critical areas, a permitted system of groundwater pumping, enforcement of technical standards, construction of waterworks and a groundwater charge system. Among these options, the most effective solution was the construction of waterworks because without such systems groundwater users would have no choice but to continue abstraction.

14.1 Introduction

Urbanization affects groundwater in various ways. Generally, as the number of people and factories in an urban area increase, some areas are used as landfills for waste disposal, whose seepage negatively impacts the quality of the underlying groundwater. Additionally, increased populations often lead to changes in land use. For example, as paddy fields and farms are converted into roads and paved with concrete, the natural recharge of groundwater is hindered. Such changes also lead to urban flooding because rainwater runs along the surface instead of being absorbed into the soil.

T. Endo (✉)
Graduate School of Life and Environmental Sciences, University of Tsukuba,
1-1-1 Tennodai, Tsukuba, Ibaraki 305-8571, Japan
e-mail: endo@envr.tsukuba.ac.jp

Among these various problems, this chapter is concerned with land subsidence. Land subsidence demands attention for several reasons. First, it is a typical urban groundwater problem that has occurred repeatedly in many Asian megacities including Tokyo, Osaka, Bangkok and Jakarta. Additionally, many Asian megacities are located in coastal areas; therefore, land subsidence often leads to flood inundation. Finally, land subsidence generally causes widespread damage to existing infrastructure such as bridges and roads.

The purpose of this chapter is to compare land subsidence policies of Osaka city (hereinafter referred to as Osaka) and Bangkok Metropolitan Administration (hereinafter referred to as Bangkok) and to deduce some lessons from these two cases. To date, many studies have been conducted to evaluate countermeasures against land subsidence in Osaka (Osaka City Waterworks Bureau 1966, 1983; Council on Comprehensive Countermeasures Against Land Subsidence in Osaka, [hereafter referred to as Council] 1972; Kataoka 2006) and Bangkok (Asian Development Bank 1994; Babel et al. 2006; Buapeng et al. 2006). Although the policies have been explained in detail in these studies, they dealt with individual cases in such great depth that they failed to address several points.

Specifically, these studies were concerned with governmental policies against land subsidence, but made no attempt to justify governmental intervention. As discussed below, governmental intervention is an effective tool, but not the only method available to solve the problem of subsidence. Previous studies did not provide clear explanations of why government policies are needed to address the problem of subsidence. Second, the studies that have been conducted to date have not clarified what types of policies are effective against land subsidence. Because countermeasures used in cities are not always the same, a case study that deals with one city in detail will not necessarily be useful for other areas.

Regarding the first shortcoming, a framework of "social dilemma" will be introduced in the second section. Specifically, land subsidence can be considered to be a social dilemma, which explains why it is difficult for private companies or local volunteers to solve the problem without governmental intervention. To address the second shortcoming, a comparative analysis in which the concept of social dilemma is the focus will be presented in the third section. Finally, in the fourth section policy lessons will be deduced from the comparative study. These lessons will expand the possible solutions to potential land subsidence in cities other than Osaka and Bangkok.

14.2 Social Dilemma

Land subsidence is often regarded as a public hazard in that its negative influence impacts many people in which it occurs. Thus prevention or mitigation of land subsidence benefits all of the residents of an area. Although it is easy to assume that individuals connected by the common benefits associated with the prevention of land subsidence would promote its prevention voluntarily, this is not always the case. If so, land subsidence would not take place repeatedly in different cities again and again. Why do people fail to prevent or stop the problem?

Groundwater is often less expensive than surface water because it does not require construction of large dams such as those required for surface water development. However, even though groundwater is cheaper, it is not free. Time and energy are required for well-digging and pumping groundwater, and there are often costs for energy associated with these activities. These are private costs in that they accrue exclusively to the resource user in response to the use of groundwater.

There are also social costs associated with groundwater pumping. For example, when individual A pumps groundwater from their own well, the overall groundwater level will decline if it is not recharged by precipitation. Although this change may be very small, the activity of individual A harms neighboring groundwater users (B, C, D...). This is because they must now pump the groundwater from deeper levels or be forced to dig another well, which increases their costs. Moreover, if the decline in groundwater leads to land subsidence, it can exacerbate the flood risk for the entire area. These additional costs are social costs because they are incurred by the neighboring groundwater users in addition to user A.

Next, let us suppose that not only A, but also B, C and D pump groundwater. In this situation, the users cause harm to each other. Although the damage caused by an individual user may be very small, when it accumulates, it results in a heavy decline in groundwater level or severe land subsidence that negatively affects the entire community. Conversely, if everyone finds alternative water sources and stops using the groundwater, they provide beneficial effects to each other. Again, although the benefit contributed by each individual is very small, the accumulation will reveal itself as restoration of the groundwater level and mitigation of flood risk for the entire area. If the value of the common benefit is bigger than the total costs of creating an alternative water supply, stopping groundwater pumping pays from a social point of view.

However, individuals may not be motivated to stop pumping groundwater because each user only obtains a small portion of the benefits genetated by doing so, even though this prevents land subsidence and mitigates the risk of flood. Additionally, finding alternative water supplies has costs, especially when groundwater is inexpensive. Accordingly, if the costs associated with not pumping groundwater outweigh the benefits each individual user can really get, stopping to use groundwater does not pay from each contributor's point of view, even though it pays from a social point of view.

If all other conditions are equal, this phenomenon will strengthen as the number of groundwater users increases. This is because, as the size of the group increases, each individual's decision to stop using groundwater has less effect. In such cases, each groundwater user will have the incentive to be a free-rider who enjoys the benefits generated by the contribution of others without making any contribution themselves. As a result, even though mitigation of land subsidence may benefit the entire community, nobody will cooperate toward realization of this goal.

This is called social dilemma or collective action problem (Olson 1965; Dawes 1975). Specifically, this situation represents a disparity between individual rationality and social rationality. Generally speaking, it is likely that people connected by a common goal will cooperate automatically. However, the theory of social dilemma shows that this is not always the case. The reason for this phenomenon is that the benefit is common, which causes all individuals in the group to expect others to contribute so that they will not have to do so. In this situation, "Everybody's business is nobody's business" is indeed true.

The framework of social dilemma can be a basis for comparative analysis of land subsidence policy in Osaka and Bangkok. In both cases, the problem was solved not by voluntary action of residents and factories, but by governmental regulations. Government is defined as an authoritative social apparatus of coercion and stimulus that can induce someone to do something by imposing penalties or rewards (Von Mises 1962). Coercion is one method of changing the pattern of groundwater use. In the next section, government intervention and alteration of groundwater use in Osaka and Bangkok will be discussed.

14.3   Countermeasures Against Land Subsidence: Osaka and Bangkok

14.3.1   A Case Study: Osaka

14.3.1.1   Land Subsidence in Osaka

Osaka has been the largest city in western part of Japan and has a population of 2.6 million in the jurisdiction of 222 km^2 as of 2008 (Osaka City 2009). The average annual rainfall from 1971 to 2000 is about 1,300 mm (Japan Meteorological Agency official website). Osaka is situated at the mouth of the Yodo River of which tributaries have been used as navigation channels. With geographical advantages for trade, Osaka began to develop as an industrial and commercial city. Both surface water and groundwater were used by many factories for their operations. As modern technologies were imported and production level increased, demand for water, especially demand for groundwater, also increased rapidly (Osaka City Waterworks Bureau 1966).

Land subsidence in western Osaka was first reported in the late 1920s by the Department of Land Survey in the Japanese Imperial Army. At the time, there were various opinions about the cause of land subsidence. However, during the Second World War, an increasing number of factories closed, and these closures corresponded with a cessation of land subsidence, which revealed that excessive groundwater pumping by factories was the main cause of land subsidence in Osaka. This opinion was confirmed when land subsidence resumed once again in western coastal areas after the Second World War as economic activities increased (Osaka City Waterworks Bureau 1966; Council 1972).

At the time, primary groundwater users in Osaka were factories and commercial buildings. While groundwater was mainly used for cooling in the industrial sector,

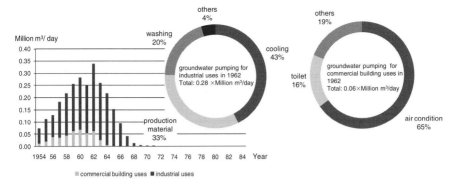

Fig. 14.1 Groundwater pumping in Osaka and its purposes. Source: Osaka City Waterworks Bureau (1966, 1983), Council on Comprehensive Countermeasures against Land Subsidence in Osaka (1972, 1985)

it was primarily used for air conditioning in the commercial sector. For example, when the volume of groundwater abstraction peaked in 1962, 43% of groundwater in the industrial sector was used for cooling purposes and 65% of groundwater in the commercial sector was used for air conditioning (Fig. 14.1). Later, the main water supply for industry changed from groundwater to surface water and recycled water. This change was possible because the water used by industry for temperature control purposes was not required to have a high quality (Kurosawa et al. 1962).

14.3.1.2 Industrial Water Law

Serious attention was first paid to land subsidence after the Muroto typhoon of 1934, which caused serious high tide damage in the Osaka area (Osaka City Waterworks Bureau 1966; Council 1972). Additionally, the government of Osaka enacted a local ordinance on civil engineering construction to regulate groundwater pumping as early as 1948. However, this act was not executed because groundwater was believed to be privately owned and there was no alternative water supply (Osaka City Waterworks Bureau 1966). Nevertheless, the government of Osaka recognized that switching from groundwater to surface water was essential to stop land subsidence (Osaka City Waterworks Bureau 1966). In 1950, the Osaka area was hit by another typhoon (Typhoon Jane), which triggered construction of an industrial water supply system that began to operate in 1955 (Osaka City Waterworks Bureau 1966, 1983).

Osaka was not the only city in Japan that suffered from land subsidence. Indeed, economic development after the Second World War was often accompanied by excessive groundwater pumping, which resulted in similar problems occurring throughout the country. As a result, the Japanese government enacted the industrial water law in 1956. The purpose of this law was to regulate groundwater pumping for industrial uses and to promote the development of industrial waterworks systems (Osaka City Waterworks Bureau 1966). Under this law, some areas were classified as "designated area" by cabinet order and every factory owner who

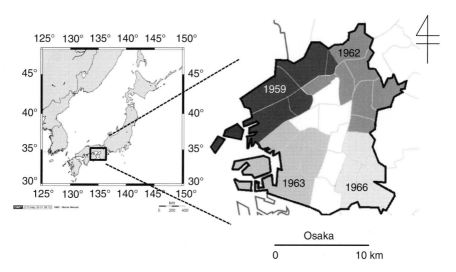

Fig. 14.2 Designated areas in Osaka under industrial water law and the year of execution. Source: Osaka City Waterworks Bureau (1983)

intended to pump groundwater in such locations was required to obtain permission from the Minister of International Trade and Industry. Osaka was classified as a designated area on December 4, 1958 and installation of wells that did not meet the requirements of the standard have not been allowed since January 4, 1959 (Osaka Waterworks 1966) (Fig. 14.2). Although conversion of the water supply is an effective method of stopping land subsidence, it will not be successful if the price of the water supplied by the industrial waterworks is high. Therefore, a national subsidy system was established by the industrial water law to help reduce the price of water supplied by industrial waterworks (Kurosawa et al. 1962).

This law was the first attempt at regulation of groundwater pumping in Japan, but there were some problems. For example, the objective of this law was limited to groundwater pumping for industrial uses. As a result, groundwater pumping for air conditioning and toilets in commercial buildings was not regulated. Moreover, creation of designated area had to meet one of the following conditions: (1) land subsidence must have already been observed in the concerned area; (2) industrial waterworks must have already been constructed in the area or scheduled to be constructed within a year. Based on these conditions, the industrial waterworks law could not prevent land subsidence. Finally, the objective of the regulation was limited to newly installed wells; therefore, existing wells were not subject to the regulation (Osaka City Waterworks Bureau 1966, 1983; Endo et al. 1975; Yanagi 2002).

14.3.1.3 Revised Industrial Water Law and Building Water Law

Land subsidence was again regarded as an urgent problem after the second Muroto typhoon hit Osaka in 1961 (Fig. 14.3). After the typhoon, the governments of

Fig. 14.3 Inundation caused by the second Muroto Typhoon. Source: Osaka Urban Engineering Information Center

Osaka Prefecture and city and business organizations appealed to the national government to create new regulations. As a result, a revised industrial water law and building water law were enacted in 1962 (Osaka City Waterworks Bureau 1966).

The revised law included revisions to the original law. First, stopping land subsidence was ranked as the top priority of the revised law (Osaka City Waterworks Bureau 1983) Second, the guidelines for the installation of wells became so strictly regulated that it was nearly impossible for new wells to be built. Finally, regulations regarding existing wells were strengthened. Specifically, existing wells in designated areas in which industrial waterworks had already been constructed that did not fit the standard set by the government were required to be removed within 1 year. These changes promoted conversion of the water supply from groundwater to surface water (Osaka City Waterworks Bureau 1983).

The purpose of the building water law was to regulate groundwater pumping for building uses such as air conditioning and toilets in locations that had been specified as a designated area by cabinet order. The building water law differed from the industrial water law in that an area could be classified as a designated area even if industrial waterworks were not constructed there. All of Osaka was specified as a designated area on August 24, 1962 (Osaka City Waterworks Bureau 1966; Yanagi 2002) (Fig. 14.4).

The standards for the installation of new wells in the building water law were similar to those in the revised industrial water law. This made construction of new wells nearly impossible (Osaka City Waterworks Bureau 1983). Additionally, the regulation covered newly constructed wells and existing wells, with existing wells being abolished after a moratorium if they did not meet the standard requirements. This moratorium ended on August 31, 1964 and groundwater abstraction for building uses was prohibited throughout Osaka (Osaka City Waterworks Bureau 1966).

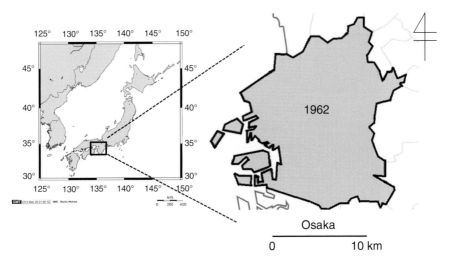

Fig. 14.4 Designated area in Osaka under Building Water Law and the year of designation. Source: Osaka City Waterworks Bureau (1966), Yanagi (2002)

In sum, the expansion of industrial waterworks and legal regulations prompted conversion of water supply from groundwater to surface water in Osaka, which resolved the land subsidence problem (Fig. 14.5).

14.3.2   A Case Study: Bangkok

14.3.2.1   Land Subsidence in Bangkok

Bangkok, which is the political and economic center of Thailand, comprises 1,569 km² and has a population of 5.7 million as of 2005 (Bangkok Metropolitan Administration, official website). Bangkok forms a huge urban area called Bangkok Metropolitan Area with its adjacent provinces including Nonthaburi, Pathumthani, Samut Prakan, Samut Sakhon, Nakhon Pathom (Tasaka 1998; Hashimoto 1998). But, in this paper, Ayutthaya province is added to those provinces for the sake of convenience (Fig. 14.6).

Bangkok is located in climatic zone of Monsoon Asia. The annual average rainfall is 1,190 mm and relatively high precipitation can be expected as compared to other dry areas. However, in Chao Phraya basin where Bangkok is situated, the river run-off concentrates during the wet season and the river flow is not frequently sufficient to meet the demand in dry season. Groundwater development has been one of the solutions to cope with water shortage problem mainly in dry season (MWA (Metropolitan Waterworks Authority) 1994; Das Gupta 2001; Das Gupta and Babel 2005).

Volume of water supply by industrial waterworks
Volume of groundwater pumping for industrial use
(Million m3/day)

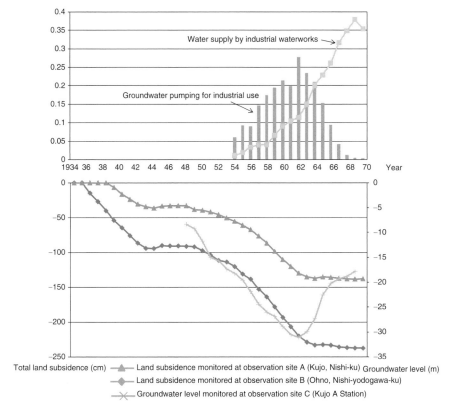

Fig. 14.5 Conversion of industrial water supply in Osaka. Source: Osaka City General Planning Bureau (1965, 1966, 1967, 1968, 1970), Osaka City Waterworks Bureau (1966, 1983), Council on Comprehensive Countermeasures against Land Subsidence in Osaka (1972, 1985), Osaka City Environment and Health Bureau (1973)

Intensive groundwater use in Bangkok dates back to 1954, when it was used to supplement the public water supply (Das Gupta and Babel 2005). Groundwater use increased dramatically due to the lower cost of development and the availability near its place of utilization (Das Gupta 2001). In 1970s land subsidence in Bangkok was first quantitatively estimated (Das Gupta and Babel 2005). Additionally, an increased risk of flood was reported after a comprehensive survey of groundwater conducted from 1978 to 1981 that revealed rapid land subsidence was occurring in Eastern and South-Eastern part of Bangkok. The city is located on the bank of the Chao Phraya River and its elevation is only 0.5–1.5 m above mean sea level. Therefore, it was assumed that land subsidence would cause serious flood and high tide risk (Asian Development Bank 1994; Ramnarong 1999; Buapeng et al. 2006).

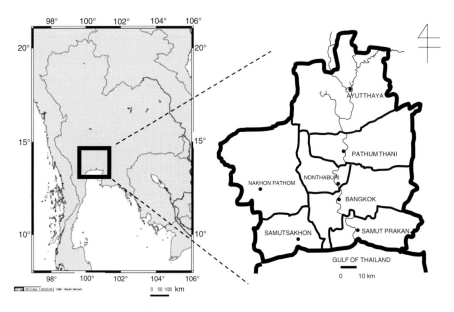

Fig. 14.6 Bangkok and the adjacent provinces

14.3.2.2 Groundwater Act

In 1977, the groundwater act was established and public regulation of groundwater pumping was introduced. The purpose of this act was to regulate groundwater pumping by the private sector in Bangkok Metropolitan Area (Ramnarong 1999; Das Gupta and Babel 2005). This rule introduced a permitted abstraction system in areas that were designated as "Groundwater Area" within Bangkok Metropolitan Area. According to this rule, anyone who wants to drill a well or inject wastewater into the ground in Groundwater Area must obtain permission (JICA and Kokusai Kogyo Co., Ltd 1995; Ramnarong 1999).

Subsequently, the Cabinet Resolution on Mitigation of Groundwater Crisis and Land Subsidence in the Bangkok Metropolis was generated in 1983. This is a long term plan for 1983–2000 that was designed to reduce groundwater pumping in areas newly designated as "Critical Zone" (Ramnarong 1999; Buapeng et al. 2006) (Figs. 14.7 and 14.8). Regulations in Critical Zone are severer than the ones in Groundwater Area in the following points. First, the total volume of groundwater that can be pumped is capped. Therefore, an application for new well is not always permitted. Second, as noted later, groundwater charge system has been introduced in Groundwater Area. In addition to this, another system called groundwater conservation charge has also been applied to Critical Zones, regardless of whether water supply of waterworks is available or not in the concerned areas (Interview 2009a; 2009b).

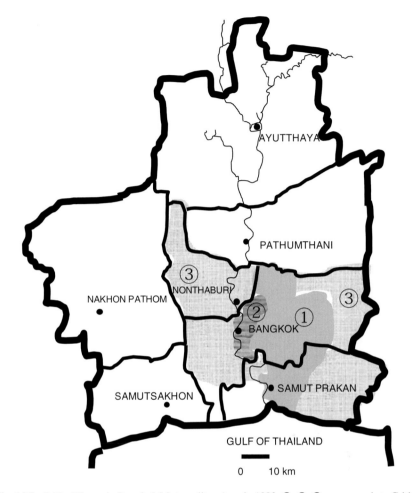

Fig. 14.7 Critical Zones in Bangkok Metropolitan Area in 1983. ①, ②, ③ corresponds to Critical Zone 1, 2, 3 respectively. The observed annual rate of land subsidence is high in order of ①, ②, ③. Source: Buapeng et al. (2006)

Critical Zones were classified into three groups depending on their subsidence rate. In Critical Zones 1 and 2, where land subsidence was relatively severe, groundwater abstraction by the Metropolitan Waterworks Authority (hereinafter referred to as MWA) for public water supply was to be reduced to zero by the end of 1987 (Ramnarong 1999). Moreover, well installation and existing wells were also prohibited or abolished where the MWA waterworks were available (Asian Development Bank 1994). Finally, this resolution included countermeasures against groundwater pumping in the private sector and planned to reduce it from 1988, when the construction of waterworks by the MWA was expected to be completed (Ramnarong 1999).

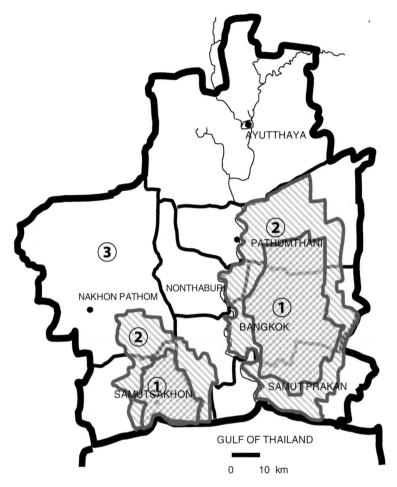

Fig. 14.8 Critical Zones in Bangkok Metropolitan Area in 1995. ①, ②, ③ corresponds to Critical Zone 1, 2, 3 respectively. The observed annual rate of land subsidence is high in order of ①, ②, ③. Source: Buapeng et al. (2006)

14.3.2.3   Waterworks in Bangkok Metropolitan Area

The waterworks in Bangkok are currently operated by the MWA, which is a state enterprise under control of the Ministry of the Interior that is responsible for supplying industrial and domestic water within Bangkok, Nonthanburi and Samut Prakan. The MWA was established in 1967 by integration of separate waterworks in the aforementioned provinces. The main water supply for the MWA is the Chao Phraya River, but it also uses groundwater to supplement the surface water. In 1970, the volume of groundwater pumping by the MWA accounted for almost half of the total abstraction in Bangkok (Babel et al. 2006), but this was gradually reduced to prevent land subsidence. Although the waterworks have expanded, groundwater pumping increased in the early 1990s because the expansion of the waterworks

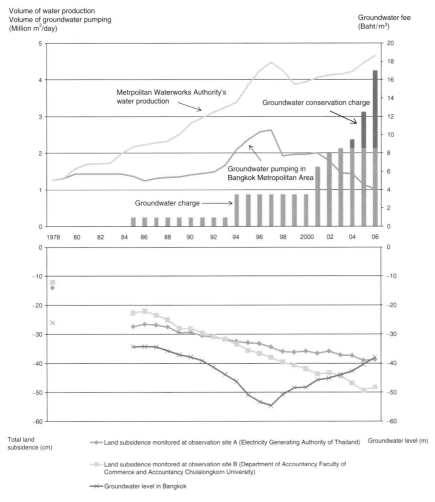

Fig. 14.9 Conversion of water supply in Bangkok. Source: Data provided by Metropolitan Waterworks Authority and Department of Groundwater Resources, Ministry of Natural Resources and Environment, Thailand

could not catch up with the rapid population increase (Phien-wej et al. 2006). However, as the water supply from the waterworks has kept increasing, groundwater abstraction has decreased (Fig. 14.9).

14.3.2.4   Groundwater Charge System

The groundwater charge system is a countermeasure that is unique to Bangkok. As mentioned before, if everyone is free to pump, excessive abstraction often occurs. Therefore, the groundwater charge system was implemented to reduce abstraction by increasing the cost of pumping. However, the effectiveness of this system depends on the price elasticity of demand and the existence of alternative water sources.

The groundwater charge system is based on the groundwater act of 1977. The system has been in place since 1985 to regulate private groundwater users in Groundwater Areas comprised of six provinces: Bangkok, Nonthaburi, Pathumthani, Ayuthaya, Samut Prakan, and in parts of Samut Sakhon (Babel et al. 2006). Initially, the price was set at 1 bahts/m^3; however, this rate was increased to 3.5 bahts/m^3 in 1994. Additionally, the charge was further increased to 8.5 bahts/m^3 between July, 2000 and April, 2003 (Ramnarong 1999; Babel et al. 2006; Buapeng et al. 2006). Despite these increases, groundwater pumping did not stop because the cost was still less than the tariff on water supplied by MWA (Das Gupta and Babel 2005).

In 2004, an additional charge system known as the groundwater conservation charge was introduced. This system charges every groundwater user in the critical zone. The groundwater fee was first set at 1 baht/m^3 in September, 2004 and then gradually increased to 8.5 bahts/m^3 in 2006. As a result, the total groundwater fee reached 17 bahts/m^3 in 2006. Because the price of tap water provided by MWA has been 15.81 bahts/m^3 since December, 1999 (a case business enterprises or governmental organizations use more than 200 m^3 of water), groundwater is now more expensive than surface water, which should induce groundwater users to change their water supply (Babel et al. 2006; Buapeng et al. 2006; MWA 2009) (Fig. 14.9).

14.4 Lessons

14.4.1 Treating Groundwater as a Unit

The first lesson that can be learned from these cases is the importance of treating groundwater as a unit. In other words, artificial boundaries may have negative impacts on groundwater management. As mentioned before, industrial water law was applied to stop land subsidence in Osaka. However, the objective of the regulation was limited to groundwater pumping for industrial uses. As a result, groundwater pumping for commercial buildings was not regulated, which limited the effectiveness of the water law. This was caused by drawing artificial lines such as "groundwater for industrial uses" and "groundwater for commercial building uses." Accordingly, these results indicate that the unit of groundwater management should be decided with attention to the whole groundwater flow system.

14.4.2 Existing Wells and Newly Built Wells

The distinction between existing wells and newly built wells is another example of an artificial boundary. The process of regulation began by restricting the installation of wells in both cases. In the case of Osaka, the industrial water law prohibited the construction of new wells in designated areas if the wells did not satisfy the standard. In the case of Bangkok, a permitted abstraction system was introduced by the groundwater act of 1977. Although the limitation of new wells had some positive effects, they were insufficient to prevent excessive abstraction. This is because groundwater is a common resource for existing and potential users. Without regulating the behavior of the former, groundwater pumping will not stop. Accordingly, the industrial water law in Osaka case was revised to strengthen the regulation of existing wells and abolish wells that did not meet the standard set by the government. Similarly, in Bangkok, the 1983 resolution gradually abolished existing wells that were located in areas in which MWA waterworks have already been constructed.

14.4.3   Geographical Distinction of Groundwater

The third example of an artificial boundary is geographical distinction. In both Osaka and Bangkok, regulation started in areas in which severe land subsidence had been observed. However, to avoid regulation, users could move to areas in which the groundwater is less stringently regulated. Should this occur, land subsidence may simply shift from one area to another. Indeed, this is likely to happen, especially when the regulated area is much smaller than the natural groundwater flow system. Accordingly, it is desirable that areas with regulation coincide with the natural groundwater flow system. However, it is difficult to understand the structure of the groundwater flow system precisely. Therefore, the only practical way to address this issue is to change the areas that are regulated in accordance with the movement of the area of subsidence.

14.4.4   Groundwater Charge System

Groundwater charge is a method of controlling groundwater pumping by increasing the cost of abstraction. This kind of policy can play a role in reduction of groundwater pumping, but it is not perfect. First, when groundwater is the sole water source, it will be abstracted regardless of the price. Conversely, when there is an alternative source of water, the groundwater fee must be higher than the fee for the alternative water to induce groundwater users to change supply sources. Second, it is legally and politically difficult for this system to be introduced where groundwater is regarded as a part of land ownership. Because a private landowner can determine the use or non-use of underlying groundwater in such a case, s/he is not likely to approve an institution that interferes with the decision-making. This explains a reason why groundwater charge system was used only in Bangkok. While groundwater is basically considered as a part of land above in Japanese legal system, the ownership of all water resources is vested in the State in Thai legal system (Miyazaki 2006; Economic Commission for Asia and the Far

East 1967). Of course, it does not mean that groundwater charge system can always be implemented within such a legal system as can be seen in Thailand, but it can be a foundation of the policy.

14.4.5 Alternative Source of Water Supply

The most effective method to stop groundwater pumping is to create an alternative water supply. In Osaka and Bangkok, waterworks were constructed to create an alternative water source. Practically speaking, economic and technical regulation may work to some degree, but it is difficult to reduce groundwater abstraction without providing another source of water. However, the construction of such systems does not always lead to a reduction in groundwater abstraction. The price of water supplied by the waterworks must be kept lower than the cost of groundwater pumping. In Osaka, this was accomplished through a governmental subsidy for industrial water, while it was accomplished mainly by increasing the groundwater charge in Bangkok.

14.5 Conclusion

The purpose of this chapter was to compare land subsidence policies in Osaka and Bangkok and provide some lessons from these two cases. Land subsidence can be regarded as an example of a social dilemma. Prevention of land subsidence is a common benefit for people who suffer flood or high tide risk. However, a group of people that share such a common benefit often fail to realize it voluntarily because each individual has an incentive to be a free-rider, who enjoys the benefit generated by the contribution of others without making any contribution themselves. Governmental intervention is an effective way to address this situation, because the government can force groundwater users to change their behavior. Indeed, this was why government intervention in Osaka and Bangkok was implemented to address the land subsidence problem. Various policies were implemented by the governments of each city, including designation of critical areas, development of a permitted system of groundwater pumping, enforcement of technical standards, construction of waterworks and implementation of a groundwater charge system. Among these options, the most effective solution was the construction of waterworks because without such systems groundwater users would have no choice but to continue abstraction.

Acknowledgments This work was financially supported by the Human Impacts on Urban Subsurface Environment project organized by the Research Institute for Humanity and Nature, Kyoto Japan. I also thank Mr. Anirut Ladawadee, Mr. Apichari Janthien, Dr.Oranuj Lorphensri, Mr. Rittikrai Bkavabhutanonda Na Mahasarakham and Mr. Wacharamedha Chandabimba

(Department of Groundwater Resources, Ministry of Natural Resources and Environment, Thailand) and Ms. Sonthaya Sinthuyont (Metropolitan Waterworks Authority, Thailand) for providing governmental documents. Finally, I wish to express my gratitude to Ms. Nuengnam Navaboonniyom and Mr. Hanchai Sawangned for their assistance in translation. The views presented here are those of the author and should in no way be attributed to Thai government or Metropolitan Waterworks Authority. Responsibility for the text (with any surviving errors) rests entirely on the author.

References

Asian Development Bank (1994) Managing water resources to meet megacity needs. In: Proceedings of the regional consultation, Manila, 24–27 August 1993. Asian Development Bank, Manila

Babel MS, Das Gupta A, Domingo NDS (2006) Groundwater resource management in Bangkok. In: Institute for Global Environmental Strategies (ed) Sustainable groundwater management in Asian cities. Institute for Global Environmental Strategies, Hayama, pp 71–80

Bangkok Metropolitan Administration official website. http://203.155.220.238/en/bt-geography.php. Accessed 29 May 2010

Buapeng S, Lorphensri O, Ladawadee A (2006) Groundwater situation and land subsidence mitigation in Bangkok and its vicinity. In: Proceedings of the 3rd international conference on hydrology and water resources in Asia Pacific Region (APHW 2006)

Council on Comprehensive Countermeasures against Land Subsidence in Osaka (1972) A history of countermeasures against land subsidence in Osaka (in Japanese)

Council on Comprehensive Countermeasures against Land Subsidence in Osaka (1985) Current status of land subsidence in Osaka (in Japanese)

Das Gupta A (2001) Challenges and opportunities for water resources management in Southeast Asia. Hydrol Sci 46:923–935

Das Gupta A, Babel MS (2005) Challenge for sustainable management of groundwater use in Bangkok, Thailand. Int J Water Resour Dev 21:453–464

Dawes R (1975) Formal models of dilemmas in social decision making. In: Kaplan MF, Schwartz S (eds) Human judgement and decision processes. Academic, New York, pp 87–107

Economic Commission for Asia and the Far East (1967) Water legislation in Asia and the Far East (Water Resources Series No.31). United Nations

Endo H, Ogawa I, Kanazawa Y, Shiono H, Takahashi Y (1975) On groundwater law. Jurist 584:16–42 (in Japanese)

Hashimoto S (1998) Public administration, public finance and development policy in urban area. In: Tasaka T (ed) Asian mega-cities, Bangkok. Nihon Hyoronsha, Tokyo, pp 281–304 (in Japanese)

Interview (2009a) Department of Groundwater Resources, Ministry of Natural Resources and Environment, Thailand, September 7, 2009

Interview (2009b) Department of Groundwater Resources, Ministry of Natural Resources and Environment, Thailand, December 21, 2009

Japan Meteorological Agency official website. http://www.data.jma.go.jp/obd/stats/etrn/index.php. Accessed 28 May 2010

JICA (Japan International Cooperation Agency), Kokusai Kogyo Co., Ltd (1995) Final report of the study on management of groundwater and land subsidence in the Bangkok metropolitan area and its vicinity

Kataoka Y (2006) Towards sustainable groundwater management in Asian cities – lessons from Osaka. Int Rev Environ Strateg 6:269–290

Kurosawa S, Nagaoka O, Murashita T, Kurata N, Sugawara K (1962) Industrial water resource. Chijinshokan, Tokyo (in Japanese)

Miyazaki A (2006) Landownership and groundwater law with special focus on the legal characteristics of groundwater. In: Urban and land use (essays in honor of Professor Inamoto Yonosuke), pp 47–73. Nihon Hyoronsha, Tokyo (in Japanese)

MWA (Metropolitan Waterworks Authority) (1994) 1993 Annual report

MWA (Metropolitan Waterworks Authority) (2009) 2008 Annual report

Olson M (1965) The logic of collective action, public goods and the theory of groups, sixteenth printings. Harvard University Press, Cambridge

Osaka City (2009) Statistics of Osaka City in 2008 (in Japanese)

Osaka City Department of Environment and Health (1973) Land subsidence in 1973 (in Japanese)

Osaka City General Planning Bureau (1965) A document on land subsidence in 1965 (in Japanese)

Osaka City General Planning Bureau (1966) A research on countermeasures against extensive public hazard with special reference to land subsidence (in Japanese)

Osaka City General Planning Bureau (1967) A document on land subsidence in 1967 (in Japanese)

Osaka City General Planning Bureau (1968) A document on land subsidence in 1968 (in Japanese)

Osaka City General Planning Bureau (1970) A document on land subsidence in 1970 (in Japanese)

Osaka City Waterworks Bureau (1966) A history of Osaka city industrial waterworks (in Japanese)

Osaka City Waterworks Bureau (1983) A history of Osaka city industrial waterworks (in Japanese)

Phien-wej N, Giao PH, Nutalaya P (2006) Land subsidence in Bangkok, Thailand. Eng Geol 82:187–201

Ramnarong V (1999) Evaluation of groundwater management in Bangkok: positive and negative. In: Chilton J (ed) Groundwater in the urban environment: selected city profiles. Taylor & Francis, New York, pp 51–62

Tasaka T (ed) (1998) Asian mega-cities, Bangkok. Nihon Hyoronsha, Tokyo (in Japanese)

Von Mises L (1962) The free and prosperous commonwealth – an exposition of the ideas of classical liberalism. D.Van Nostrand Company Inc., Toronto

Yanagi K (2002) On legal system on groundwater. Groundwater Tech 44:2–8 (in Japanese)

Chapter 15
Chemical and Physical Evidences in the Groundwater Aquifer Caused by Over-Pumping of Groundwater and Their Countermeasures in the Major Asian Coastal Cities

Jun Shimada

Abstract Economic growth of urban area induces the huge water demand for the city production and this creates the groundwater related disasters in many Asian cities. This situation has started from 1960s to 1970s Japan. After, many coastal cities in South East Asia had experienced similar problems later 1980s–2000s. This was caused by over-pumping of groundwater in the urban area and the related local city governments had tried to make countermeasures to protect such groundwater disasters by adapting the groundwater pumping regulation with the help of their national government. In the case of Japan, it has good success and the dropped groundwater level has clearly recovered by the help of the worm humid hydrological condition of Japanese island. Similar recovery has confirmed in Taipei and Bangkok, but it has not yet succeeded at Jakarta.

At the planning stage of the present project, we believe that those groundwater over-pumping situations must create the forced groundwater flow in the related aquifer and the chemical information on the environmental change of those urban areas should be remained as the precise time series information along the groundwater flow line in that aquifer like the paleo-hydrology record in aquifer. For this particular purpose, we have developed the young age tracer of groundwater such as CFCs and ^{85}Kr for the recent 100 years historical data analysis. However, we have confirmed that it could not use CFCs age tracer in the urban aquifers because of man-made local CFCs contamination especially dominated at urban area. Instead, we showed that the CFCs contents at urban area can be useful to understand the induced relatively young vertical flow flux through shallower aquifers caused by the man made depressed groundwater potential under the city area. This kind of situation has been confirmed at Tokyo, Bangkok and Jakarta area and these vertical induced fluxes are much more than the natural lateral groundwater flow along the aquifer. The 3D groundwater flow simulation clearly showed this kind of induced vertical flux ('induced recharge') dominated at the groundwater depression area where was created by the over-pumping of the groundwater resources in city area.

J. Shimada (✉)
Graduate school of Science and Technology, Kumamoto University,
2-39-1 Kurokami, Kumamoto 860-8555, Japan
e-mail: jshimada@sci.kumamoto-u.ac.jp

In spite of these severe situations, our comparative study clearly shows that the potentiality of the positive groundwater recharge in the coastal Asian city aquifer. It works effectively to recover the depressed groundwater level, if the groundwater pumping regulation works perfectly with the help of infrastructure construction of the additional surface water supply system and the steady legal system for the groundwater resources.

15.1   Introduction

Most Asian coastal cities has experienced serious groundwater disasters during recent 50 years caused by the over-pumping of regional groundwater related with the abrupt demand of groundwater resources linked with the urban economic growth (Fig. 15.1). These situations have started initially at Osaka area, Japan in

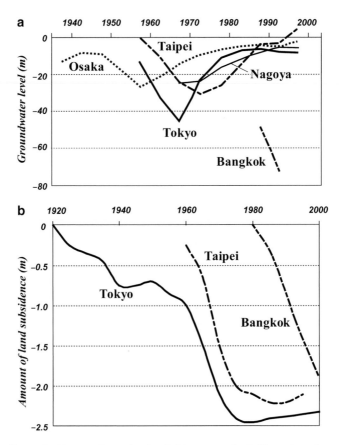

Fig. 15.1 Change in (**a**) groundwater level and (**b**) amount of subsidence in major coastal cities in Asia (Taniguchi et al. 2009)

1960s, then after Tokyo and Nagoya areas in 1970s. As these economically important Japanese cities developed on the alluvial coastal sediments, the related groundwater disasters are either huge land subsidence or salt-water intrusion in the coastal areas. As there was no unified national groundwater law in Japan, the groundwater disaster-related prefectures cooperated to establish the groundwater regulating audiences over the problem area. National governments helped to construct the alternative infrastructure of surface water supply system for the industrial and city water demands. Because of those countermeasures, the groundwater disasters in the previous Japanese three large cities has almost disappeared within 30 years periods, which is surprisingly quicker than the most hydro-geologist's prediction. This quick recovery of the depleted groundwater potential can be explained by the positive groundwater recharge potentiality of Japanese island, which belongs to the warm humid hydrological region.

A similar groundwater disaster has been expanding to the major Asian coastal cities, such as Taipei, Shanghais, Manila, Bangkok, Jakarta, etc., which grows under the framework of the delayed economical expansion of the Asian nations within the 40 years after 1970s. Some cities like Taipei and Bangkok has succeeded to control such groundwater disasters using the similar way as Japan. However, cities like Jakarta and Manila has still in the serious problem and has not found their correct way-out.

15.2   Paleo-Hydrology in Urban Aquifer

Paleo-hydrology is the idea to use the chemical (including isotopic) information in the groundwater aquifer, as climatic or hydrological proxy along the re-constructed time-series record with the help of groundwater age tracer (Fig. 15.2). It has developed in 1990s mostly in the continental large aquifer of relatively small present recharge in the dry climatic condition. It can retrieve 50,000 to one million-year paleo-record in the inland part of the present earth.

Northern China Plain (NCP) is the representative crop production area of present China where cultivate maize in summer and wheat in winter to support the huge food demand of coastal China. For those crop productions, the intensive irrigation was introduced after 1950s Chinese independence and Yellow river water and the groundwater in the NCP aquifer was heavily used. The over pumping of NCP aquifer creates the huge depletion of groundwater potential nearly 70 m down at maximum. Northern China Plane is relatively dry hydrological condition with 600 mm annual precipitation and this creates little recharge to the NCP aquifer. However, the previous man made over-pumping from NCP aquifer caused the accelerated and induced recharge in the recharge area of NCP aquifer which has been confirmed by the groundwater age tracer of ^3H and ^{14}C (Shimada et al. 2002).

Similar situation should be created in the recharge area of urban aquifers of coastal Asian area, where has experienced serious over-pumping problem shown in the former section. Compare to the NCP aquifer, the local aquifers of the urban area

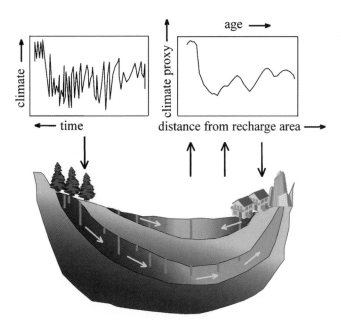

Fig. 15.2 Concept of paleo-hydrology (paleo-information extracted from groundwater aquifer)

of the coastal Asia are relatively small, but the potential recharge rate is relatively large because of their hydrological setting. This means that quite young recent recharge component should be induced into that urban aquifer by the accelerated recharge caused by the over-pumping at the city areas. To clarify this, we need to develop reliable young age tracer of groundwater instead of ^{3}H, which has lost young age tracer activity after the 50 years absence of bomb tritium-peak radioactivity. For this purpose, we start to set-up two groundwater-aging techniques: CFCs and ^{85}Kr, which are believed to be ideal young groundwater age tracer after 1980s. CFCs which has composed by three components; CFC 11, CFC 12, and CFC 113, are the man made organic gases developed 1950s and widely used as refrigeration and air conditioning or spray-gas until 1990 when was stopped to use to keep the global Ozone layer. ^{85}Kr is the man made radioactive isotopes with 10.7-year half-life and originate from reprocessing plant of used nuclear fuels; its concentration is keep increasing after 1960s. The concentration trend of both age tracers and other groundwater age tracers are shown in Fig. 15.3.

The CFCs age method has introduced to follow the USGS method through this project, but this is the first trial in Japan. ^{85}Kr age method has newly developed by this project for its on-site dissolved gas extraction system by using fiber membrane method (Ohota et al. 2009), and for the radioactivity measurement by low level LSC system (Momoshima et al. 2010). The CFCs age measurement system has fixed until 2006 and applied our field study. The ^{85}Kr method has completed its prototype system and start to check its applicability at some of our field sites.

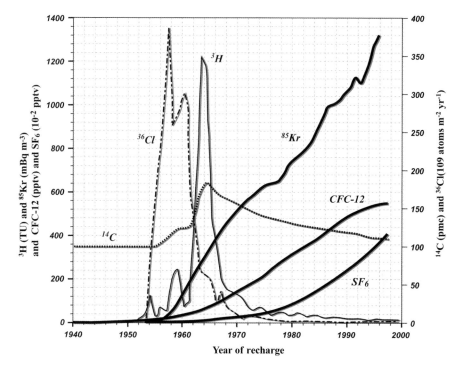

Fig. 15.3   CFC 12 and ^{85}Kr for modern groundwater age tracer

15.3   Change of Potential Distribution of Urban Aquifer

15.3.1   Kanto Plain Including Tokyo Area

Figure 15.4 shows year to year change of groundwater level for the major observation well of Tokyo area. The long-term trend of groundwater level at the observation well in Tokyo University clearly shows groundwater depression caused by the economic production of Tokyo area; it has small draw down just before WWII (1945) and little recovery after the war, then abrupt down after 1950 and has the lowest peak in 1970s. The accumulated amount of land subsidence shown as dashed line in Fig. 15.5 becomes largest in 1970s. After 1961, the inter-regional groundwater regulatory audience for the industrial groundwater use and building groundwater use was established and applied from mostly depressed area to the surrounding area until 1972. Those regulatory actions are summarized in the commentary part of Fig. 15.5.

After 1970s, the groundwater potential draw down has clearly stopped and start to recover by the effect of those groundwater regulation countermeasures. The average recovery rate is 2.0–3.5 m/yr and it is quite faster than the most hydro-geologist's prediction. This is because of the relatively higher potential groundwater recharge

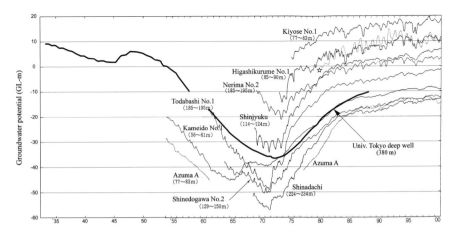

Fig. 15.4 Change of groundwater level for the major observation well of Tokyo area

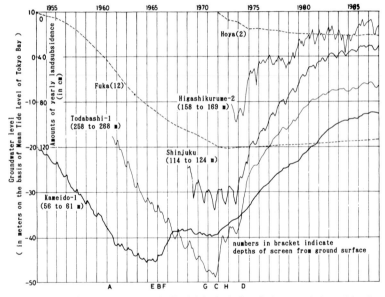

A: In January 1961, southern part of Alluvial Lowland, Koto Region, was designated by the Industrial Water Law(IWL) as a restricted area where no new wells were to be installed for industrial usage.

B: January and June 1966, pumping of groundwater for industrial usage in Koto Region was restricted by IWL.

C: In December 1971, pumping of groundwater for industrial usage in the northern part of Alluvial Lowland called Johoku Region was restricted by IWL.

D: In April 1974, pumping of groundwater for industrial usage was more reinforced in Johoku Region by IWL.

E: In July 1965, pumping of groundwater for airconditioning in Alluvial Lowland of Wards District was restricted by the Law Controlling Pumping of Groundwater for Use in Building(LCB).

F: In July 1966, pumping of groundwater for airconditioning in Terrace of Wards District was restricted by LCB.

G: In November 1970, drilling of new wells for industrial and non-drinking usages was restricted in Tama District under the Metropolitan Ordinance.

H: December 1972, extraction of water-soluable natural gas was suspended in the estuary of the Ara River by the means of purchase of the mining rights by Tokyo Metropolitan Government.

Fig. 15.5 Long term trend of groundwater level in Tokyo area and application of its regulatory countermeasures (Endo 1992)

rate under the warm humid hydrological condition of Japan. In early 1990s, the most depressed area has showed as high as 40 m recovery to 10–15 years before. However, this situation caused another groundwater disaster; floating power problem of underground structures. In the case of Tokyo and Ueno Shinkansen underground station which was constructed 30 m below ground surface, and its groundwater level was lower than 30 m below surface when it was constructed. Now the groundwater level of Tokyo and Ueno station area are around 5 m below surface, and the underground station has already deep under the groundwater level. This situation creates not only huge amount of groundwater leakage into the station tunnel to be pumped up, but also makes the huge upward floating power to the underground structure itself. Normally this kind of upward floating power can be compensated by the building structure weight itself above ground, but station structure has not such weight because of the original surface railway system just above the underground station has already existed and could not make any high rise structure above ground. To compensate this huge floating power, JR east Railway Company must put heavy weigh under the platform and also make earth anchor connecting to the deeper base rock formation (Fig. 15.6).

Although these unexpected quick recovery of groundwater potential caused such new groundwater disaster, the groundwater recovery plan itself has well succeeded. However the pumping regulation audience applied over the groundwater depression area and do not cover whole Kanto region where develops the major groundwater aquifer of Tokyo area and has the marginal recharge area of the groundwater flow toward Tokyo area. This situation caused the shifting of the groundwater depression area from Tokyo area toward northern Kanto where is no pumping regulation has established. As shown in Fig. 15.7, the depressed area in the major regional aquifer has moved toward northern inland part in recent 30 years and there happened another land subsidence problem caused by the new groundwater depression (Hayashi et al. 2009).

Fig. 15.6 Countermeasures to protect the floating power caused by the groundwater recovery

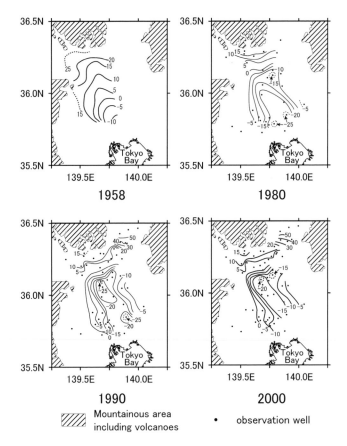

Fig. 15.7 Shifting of depressed area of groundwater potential during last 30 years (Hayashi et al. 2009)

Figure 15.8 shows the change of recharge area in Kanto plain during last 60 years estimated from the 3D groundwater flow simulation. Because of the shifting of the groundwater depression area toward outside of the pumping regulation area, the recharge area of the groundwater also moved from eastern Tokyo in 1960s, then expanded to western Tokyo in 1970s. After the effective pumping regulation at eastern Tokyo and its surrounding coastal area in 1990–2000, the previous induced recharge at eastern and western Tokyo area has gradually disappearing as shown in Fig. 15.8. Instead, after 1970s, another recharge area has start growing at Northern part of Kanto Plain and gradually expanding until now.

Figure 15.9a shows the vertical cross sectional distribution of the groundwater potential, age components (^{14}C, ^3H, CFC 12) and stable hydrogen along the dashed line from recharge area to the central depression area of Kanto plain shown in the corner index map. As shown in Fig. 15.9b, stable hydrogen having relatively

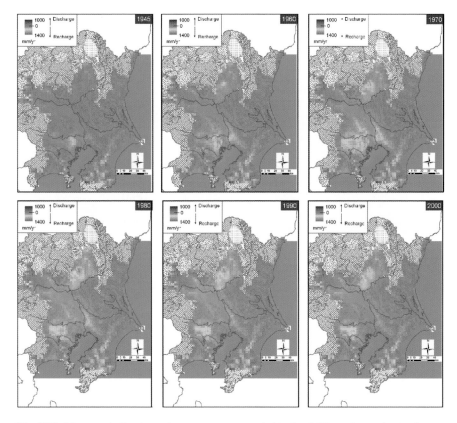

Fig. 15.8 Movement of major recharge area accompanied to the shifting of groundwater depression area caused by the application of regional groundwater regulation (Aichi and Tokunaga 2008)

depleted component, which reflect the much cooler climate than present, shows the age of last glacial maximum by apparent ^{14}C age. These old age water mostly exist in the marginal recharge area, while the central potential depressed area shows relatively young age component by relatively high CFC 12 concentration and relatively heavy stable hydrogen content as shown in Fig. 15.9b. These unexpected age distribution along the flow line from recharge area to the lower depression area means that the young shallow component has introduced to the depression part aquifer mostly from the upper shallower aquifer and the lateral flow through original aquifer is not much dominant by this depression as we expected previously.

We consider that this situation is the clear evidence of the enforced young water intrusion caused by over-pumping potential depression and we could call this as 'induced recharge'. In this case, CFC 12 content in the groundwater does not tell its exact age because of the relatively higher contamination of CFCs content in the urban area but it can be applicable as the indicator of the modern groundwater component aged as 1960–1970s.

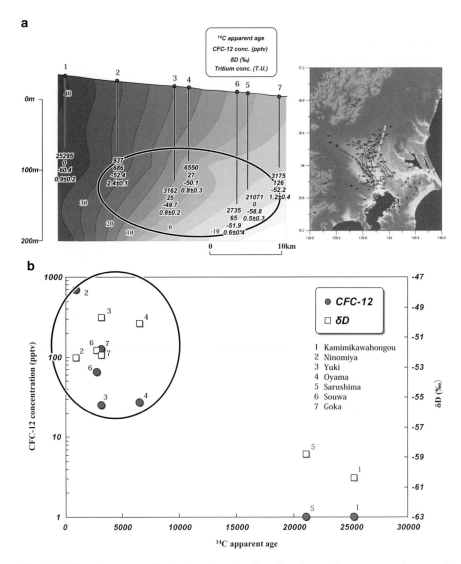

Fig. 15.9 Groundwater age distribution along the flow line from recharge area to the central depression area of Kanto plain

15.3.2 Bangkok Plain

The Bangkok Basin is filled with Quaternary sediments, and upper 600–650 m of which are subdivided conventionally into eight aquifers: the Bangkok (BK), Phra Pradaeng (or expressed as Phra Pradang; PD), Nakhon Luang (NL), Nonthaburi (NB), Sam Khok (SK), Phaya Thai (PT), Thon Buri (TB), and Pak Nam (PN) aquifers. In the southern part of Chao Phraya delta, the Bangkok aquifer is covered by the Bangkok clay (BC), while in the northern part the BC layer is nonexistent and the ground surface is covered by fluvial or alluvial deposits (Sinsakul 2000).

Although the exact year in which groundwater pumping from confined aquifers started is not known, the first large-scale exploitation of deep groundwater for public water supply began in 1954 (Das Gupta and Yappa 1982). Since then, groundwater pumping became excessive and induced land subsidence and groundwater salinization in the Bangkok metropolitan area (BMA) and Ayutthaya Province. To address this groundwater crisis, the national government took various measures, and total groundwater pumping in the BMA has reduced since the late 1990s. The piezometric levels in most wells have restored, responding to the reduction in groundwater pumping (Yamanaka et al. 2010a, b). Figure 15.10 shows the change of observed groundwater potential of NL aquifer, which is the major aquifer of Bangkok area.

Although it was nearly 30 years delay from the experience at Kanto Plain, the depression of groundwater potential was clearly caused by the over-pumping of the groundwater at the city of Bangkok and its surrounding area. This draw-down situation caused serious land subsidence in the urban Bangkok area and national government of Thailand has establish groundwater regulation law which include the groundwater charge starting from 1985 and groundwater preservation charge starting from 2003. Also Central government developed the construction of the city water supply (MWA) by surface water from early 1980 as the alternative water source to the city. As was shown in Fig. 15.11, those countermeasures work clearly and the total groundwater abstraction of the Bangkok area has started to decrease from early 2000s. After 2007, the unit groundwater abstraction cost by those charge settled 17 Bah/m^3, which is much expensive than the cost of MWA city water (16 Bah/m^3) and this works to accelerate the decrease of groundwater abstraction rate drastically.

Figure 15.12 shows the CFC 12 content along the flow line from the recharge area of plain margin to the depression part of the urban area of the Bangkok city (shown as A-B-C line in the index map) (Yamanaka et al. 2010a, b). Here again, CFCs as the modern age components of the urban area has been confirmed in the depressed area of groundwater potential, the young shallow component has been introduced to the depression part of the aquifer mostly through its upper shallow aquifer and the lateral flow through original aquifer is not much dominant by this groundwater potential depression. This situation is again the clear evidence of the enforced young water intrusion caused by groundwater over-pumping. Those 'induced recharge' is also confirmed by the 3D groundwater simulation study (Yamanaka et al. 2010a, b) shown in Fig. 15.13.

15.3.3   Jakarta Area

Jakarta, the capital city of the Republic of Indonesia, is bordered by the South China Sea in the north and has a surface area of approximately 652 km^2 under the humid tropical climate markedly influenced by monsoon with 1,500–2,500 mm of annual rainfall.

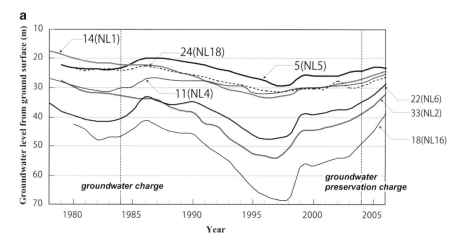

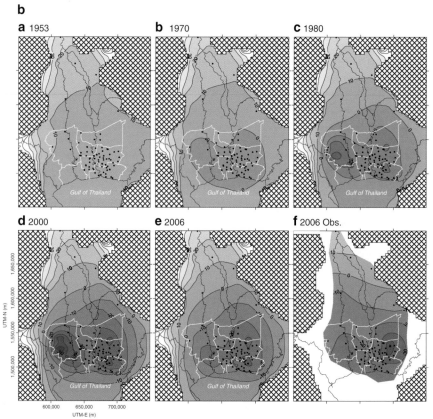

Fig. 15.10 (**a**) Long term groundwater potential record of NL aquifer, Bangkok. (**b**) Change of groundwater potential distribution of NL aquifer, Bangkok. a) to e) are model estimation using some observed data, and f) is observed distribution

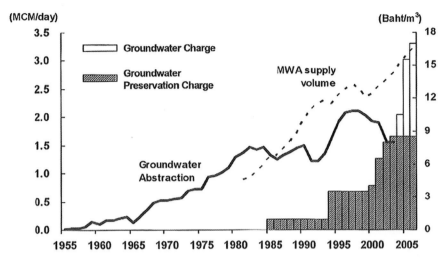

Fig. 15.11 Change in groundwater abstraction, MWA supply volume and the Groundwater charge (Babel et al. 2007)

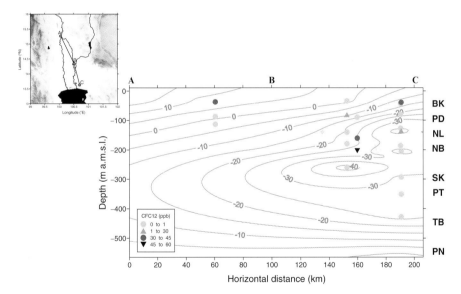

Fig. 15.12 CFCs content overlaid on the vertical groundwater potential distribution along the groundwater flow line in the Bangkok aquifers

The Jakarta Groundwater Basin, in which the city of Jakarta locates, is one of the most developed basins in Indonesia. The aquifer system in this basin is classified into five zones as follows; Zone 1 as a shallow aquifer layer composed of sandstone, conglomerate and claystone, Zone 2 and 4 as an aquiclude layer composed of claystone with sand infix, Zone 3 as deep aquifer layer composed of sandstone with infix

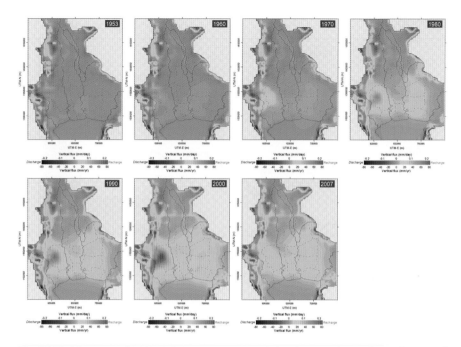

Fig. 15.13 Movement of major recharge area accompanied to the groundwater depression caused by over pumping and its countermeasure regulation

of breccias and claystone, Zone 5 as the basement of Jakarta groundwater basin composed of impermeable rock such as limestone and claystone. Components of those aquifers such as sandstone, conglomerate are connected each other and the thickness of the deep aquifer below the urban area is about 150 m.

Urbanization has obviously had a marked impact on the water demand in this area. Since only 30% of the city's potable water supply is met by surface water (Delinom et al. 2009), people are increasingly turning to the groundwater resources of the basin. Reliance on groundwater as a major water source is common and demand has been reported to reach 76% (Karen 2007). The amount of groundwater abstraction and the number of abstraction wells are shown in Fig. 15.14, which clearly shows that groundwater demand has increased in proportion to the development of the city.

Those abrupt groundwater demand increase affect seriously to the local groundwater aquifer system. Figure 15.15 shows the groundwater potential from 1985 to 2008. In 1985, the depression area of the groundwater potential was located in the northeastern part of the DKI Jakarta and the groundwater potential was −15 m. The depression area then expanded, moving toward central DKI Jakarta before moving to the northwestern part of the city in 2008 when it was decreased to −25 m. Although the Indonesian government has tried to regulate the groundwater over-pumping of this area, it has not well succeeded until now. This is probably because the development of the alternative water source by surface water is not sufficient to replace it.

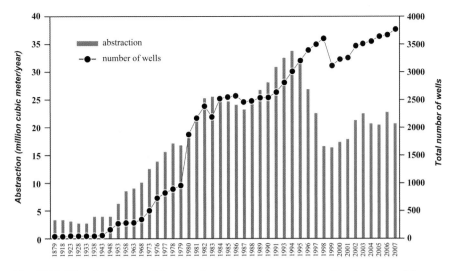

Fig. 15.14 Historical change of the groundwater abstraction amount and total number of abstraction well (Ministry of Energy and Mineral Resources, Republic of Indonesia 2007)

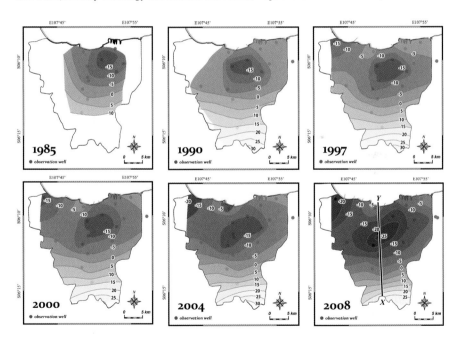

Fig. 15.15 Change of the groundwater potential distribution of Jakarata area from 1985 to 2008 (Kagabu et al. 2010)

Figure 15.16 shows the distribution of ^{14}C value (in pmC) and CFC-12 concentration along X-Y cross-section overlaid on the groundwater potential of 2008 (Kagabu et al. 2010). Even in the deeper aquifer, it has some modern ^{14}C component and which becomes higher than that of 1985 confirmed by the duplicated ^{14}C measurements

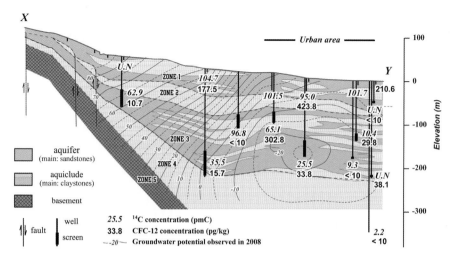

Fig. 15.16 The distribution of ¹⁴C value (in pmC) and CFC-12 concentration along X-Y cross-section overlaid on the groundwater potential of 2008 (Kagabu et al. 2010)

for those sampling points. Those 'apparent age rejuvenation' are clearly related with the groundwater drawdown caused by the over-pumping. The intrusion from surface younger groundwater component becomes nearly 50% in the total abstraction at the urban area in 2008, which has confirmed by the 3D groundwater simulation of the Jakarta area (Kagabu et al. 2010). This is also the evidence of 'induced recharge' by the over-pumping.

15.4 Change of Groundwater Recharge Caused by Groundwater Abstraction

15.4.1 Effect of Tunnel Construction to Change the Groundwater Recharge

When we construct the tunnel, the groundwater flow system in the surrounding host rock should be affected by the leakage to the tunnel. Figure 15.17 is the result of the 2D groundwater simulation case study to show the effect of tunnel construction to the groundwater system around the tunnel. The geology of the bedrock is granite and it has fixed homogeneous permeability with 10^{-6} cm/s. Before the tunnel construction (Fig. 15.17a), the groundwater recharge is 0.5 mm/day to support the stream discharge just above the proposed tunnel site which is less than average specific river discharge rate of Japan. This is probably due to the relatively low value of the fixed permeability of the studied granite rock. After the tunnel construction as shown in Fig. 15.17b, the recharge rate of 1.0 mm/day is necessary to support the stream discharge above the tunnel. If the recharge condition after the

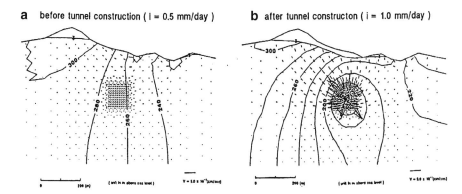

Fig. 15.17 Effect of tunnel construction to the groundwater flow and groundwater recharge (Shimada and Ishii 1986)

tunnel construction is kept at the same rate (0.5 mm/day) as before the tunnel construction, the groundwater table has been draw-down toward the tunnel and the stream flow would disappear. In order to keep the stream discharge as same as even after the tunnel construction, this study showed that the groundwater recharge rate should be changed from 0.5 to 1.0 mm/day, which is the average specific river discharge rate in Japan. Although the stream discharge has decreased somehow but still kept on running even after the tunnel construction in the actual case, this can be considered that the recharge rate to the granite rock aquifer has changed shown as the simulation study and this is the good example of the 'induced recharge' by the tunnel construction.

This kind of situation can be happen at the groundwater over-pumping city area where the 'induced groundwater recharge' could be created up to the regional maximum potential recharge rate, which is P(precipitation)–AE(actual evapotranspiration) of the concerned area.

15.4.2   Potential Groundwater Recharge in Coastal Asia

Figure 15.18 shows the hydrological setting of Asian area confirmed by the observed meteorological data. Southeastern part of continental Asia and surrounding islands countries belongs mostly to the monsoon climate with higher precipitation and moderate evapotranspiration. The residual component of the water budget, P–AE, is always positive in these area which means there is no water deficit throughout the year.

The target cities of this project, Tokyo, Osaka, Taipei, Bangkok, Jakarta, Manila, are mostly belong to this positive P–AE area as shown in Fig. 15.18, which means groundwater can be recharged up to local P–AE amount. That is why the groundwater regulation of Japanese large cities and Taipei area effectively worked and to show the quick recovery of their groundwater potential under the effective pumping regulation periods. Although Bangkok is just starting its pumping regulation, it has already shown some gradual recovery of their groundwater potential. Although

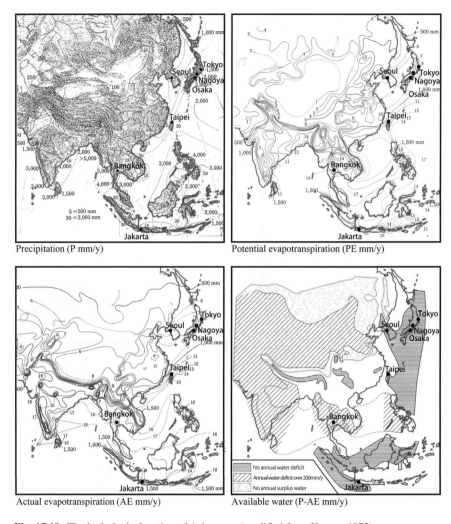

Fig. 15.18 The hydrological setting of Asian area (modified from Kayane 1972)

Jakarta is not well regulated for their groundwater use at the present stage, it has a possibility to recover under the specified groundwater regulation because of their potentially positive P–AE.

15.5 Comparison of Groundwater Development Steps in Coastal Asia

The groundwater use in the city area can be divided into several development steps depend on their economic situation. Most of Asian coastal city begin to locate along the riverside flatland because of the easy water access either for transportation or water demand for life and production use. The groundwater was developed at the inland part of the city where is less accessibility to the river water but easy access to the shallow aquifer. At this step, groundwater resources are stable, clean, cheap and easy access for the city (Step 1). Then following the economic growth of the city, water demand is also became large and the groundwater related disasters; such as land subsidence, water salinization, oxygen gas deficit accident, etc., was found significantly in the urban area (Step 2). To countermeasure these disasters, groundwater-pumping regulation has been applied to the problem area with the help of infrastructure of surface water supply as the alternative city water source (Step 3). For the good success of this groundwater pumping regulation, the legal aspect on groundwater resources is also important. In the case of coastal Asian cities, most cities belong to monsoon climatic condition area and have the potential positive groundwater recharge rate as shown in the previous section. Because of this hydrological background, the groundwater drawdown can recover very quickly if the pumping regulation works perfectly (Step 4). This situation has been experienced in the major Japanese big coastal cities like Tokyo, Nagoya and Osaka in 1970 and 1980s. However, unexpected another groundwater related disaster has appeared in the pumping regulation area, that is the floating power problem caused by the groundwater recovery explained in the previous section (Step 5). Those differences of the development steps in the coastal Asian cities are summarized in the Fig. 15.19. For the sustainable use of the city groundwater in the coastal Asia, it is important to manage the groundwater resources under the framework of hydrological circulation of the concerned area. When we have the information on the hydro-geological structure

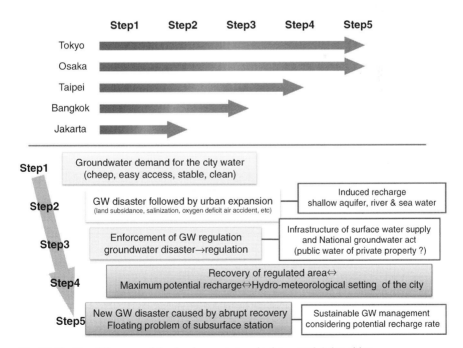

Fig. 15.19 The differences of the development steps in the coastal Asian cities

of the aquifer, its precise flow system, potential recharge area and its rate, effective pumping regulation, it is possible to promote the sustainable management of their groundwater resources in the coastal Asia.

15.6 Conclusions

In this project, we have conducted the comparative study of the coastal Asian cities for their groundwater problems. Japanese large cities has experienced severe groundwater disasters during 1960 and 1970s and succeeded to recover by the effective groundwater pumping regulations. Many Asian coastal cities like Taipei, Shanghai, Bangkok, Manila, and Jakarta have similar groundwater disasters 10 to few 10 years late after Japan. Some cities like Taipei and Bangkok has succeeded for the recovery of their groundwater potentials but many cities are still in the problem and seeking for the better solution. Although there exists the historical difference depend on the economical development steps for these cities, the groundwater problems in the coastal Asian cities shows the similar historical trends and it is said to be possible to recover based on their hydrological setting of coastal Asia. Followings are the concluding remarks of this section.

Over-pumping at urban area creates the groundwater disasters such as land subsidence, sea water intrusion, oxygen air deficit problems, etc. and also 'the induced groundwater recharge' at the depression area of groundwater potential, these include;

1. Forced recharge at the recharge area of the target aquifer
2. Squeezing from the upper and lower confining bed of the target aquifer
3. Intrusion of the shallow groundwater through the confining bed which has confirmed by the groundwater age component such as CFCs, ^{3}H and ^{14}C

Coastal Asia has the potential recharge capacity based on its hydrological setting and its maximum recharge rate should be P–AE. Because of this, it is possible to regulate and manage the groundwater resources in Coastal Asia. To succeed in the regulation, the infrastructure of surface water supply and the background of national groundwater act is key issue. Also understanding the hydro-geological setting of the local aquifer and the evaluation of the maximum recharge capacity of the area is important for the effective groundwater management. The historical groundwater disaster and its recovery experience of Japanese large cities can contribute much for the cities in coastal Asia.

References

Aichi M, Tokunaga T (2008) Estimation of the spatio-temporal change of the groundwater recharge in the Kanto Plain, Japan, from numerical simulation. In: Proceedings of IAH Toyama conference, Oct. 2008, p S23

Babel MS, Gupta AD, Sto Domingo ND, Kamalamma AG (2007) Sustainable groundwater management in Bangkok. Sustainable groundwater management in Asian Cities. Institute for Global Environmental Strategies, pp 26–43

Das Gupta A, Yappa PNDD (1982) Saltwater encroachment in an aquifer: a case study. Water Resour Res 18:546–556

Delinom RM, Assegafb A, Hasanuddin ZA, Taniguchi M, Suhermana D, Lubis RF, Yuliantoa E (2009) The contribution of human activities to subsurface environment degradation in Greater Jakarta Area, Indonesia. Sci Total Environ 407:3129–3141. doi:10.1016/j.scitotenv.2008.10.003

Endo T (1992) Confined groundwater system in Tokyo. Environ Geol Water Sci 20(1):21–34

Hayashi T, Tokunaga T, Aichi M, Shimada J, Taniguchi M (2009) Effects of human activities and urbanization on groundwater environments: as example from the aquifer system of Tokyo and the surrounding area. Sci Total Environ 407(9):3165–3172

Kagabu M, Shimada J, Nakamura T, Delinom R, Taniguchi M (2010) The groundwater age rejuvenation caused by the excessive groundwater pumping in Jakarta area, Indonesia. J Hydrol (under submission)

Karen B (2007) Trickle down? Private sector participation and the pro-poor water supply debate in Jakarta, Indonesia. Geoforum 38:855–868

Kayane I (1972) Hydrological region of Monsoon Asia (in Japanese). Science Report of Geography, Tokyo University of Education 16:33–47

Ministry of Energy and Mineral Resources, Republic of Indonesia (2007) Report of Groundwater abstraction in Jakarta 1897–2007 (in Indonesian language)

Momoshima N, Inoue F, Sugihara S, Shimada J, and Taniguchi M (2010) An improved method for ^{85}Kr analysis by liquid scintillation counting and its application to atmospheric ^{85}Kr determination. J Environ Radioact 101(8):615–621

Ohota T, Mahara T, Momoshima N, Inoue F, Shimada J, Ikawa R, Taniguchi M (2009) Separation of dissolved Kr from a water sample by means of a hollow fiber membrane. J Hydrol 376:152–158

Shimada J, Ishii T (1986) The influence on water chemistry in the crystalline bed-rock groundwater by Tsukuba tunnel construction. In: Proceedings of 8th geo-technical and geohydrological aspects of waste management, Fort Collins, USA, pp 467–476

Shimada J, Tang C, Tanaka T, Yang Y, Sakura Y, Song X, Liu C (2002) Irrigation caused groundwater drawdown beneath the North China Plain. In: Proceedings of int'l groundwater conference, Darwin, Northern Territory, Australia, May 2002, pp 1–7

Sinsakul S (2000) Late Quaternary geology of the lower Central Plain, Thailand. J Asian Earth Sci 18:415–426

Taniguchi M, Shimada J, Fukuda Y, Yamano J, Onodera S, Kaneko S, Yoshikoshi A (2009) Anthropogenic effects on the subsurface thermal and groundwater environments in Osaka, Japan and Bangkok, Thailand. Sci Total Environ 407(9):3153–3194

Yamanaka T, Shimada J, Tsujimura M, Lorphensriand O, Mikita M, Hagihara A, Onodera S (2010a) Tracing a confined groundwater flow system under the pressure of excessive groundwater use in the Lower Central Plain, Thailand. Hydrol Processes (in press)

Yamanaka T, Mikita M, Lorphensriand O, Shimada J, Kagabu M, Ikawa R, Nakamura T, Tsujimura M (2010b) Anthropogenic changes in a confined groundwater flow system in the Bangkok Basin, Thailand, Part II: How much water has been renewed? Hydrol Processes (accepted for publication)

Index

M. Taniguchi (ed.), *Groundwater and Subsurface Environments: Human Impacts*   311
in Asian Coastal Cities, DOI 10.1007/978-4-431-53904-9, © Springer 2011